Commercial Goat Production

J. M. Wilkinson
BSc, PhD, CBiol, MIBiol

and

Barbara A. Stark
BSc, PhD, CBiol, MIBiol

First published 1987

British Library
Cataloguing in Publication Data
Wilkinson, J.M.
    Commercial goat production.
    1. Goats
    I. Title   II. Stark, Barbara A.
    636.3′9   SF383

ISBN 0–632–01848–8

BSP Professional Books
Editorial offices:
Osney Mead, Oxford OX2 0EL
    (*Orders*: Tel. 0865 240201)
8 John Street, London WC1N 2ES
23 Ainslie Place, Edinburgh EH3 6AJ
52 Beacon Street, Boston
    Massachusetts 02108, USA
667 Lytton Avenue, Palo Alto
    California 94301, USA
107 Barry Street, Carlton
    Victoria 3053, Australia

Set by V & M Graphics Ltd,
    Aylesbury, Bucks
Printed and bound in Great Britain
by Billing & Sons Ltd. Worcester

Contents

Preface

This is a book about making money out of the products from herds of goats. It is about how to develop a goat herd as a business activity rather than as a hobby. It is about technical efficiency, product quality and successful marketing of goat milk, meat and fibre.

Interest in commercial goat production continues to grow as shops increasingly carry a wider range of dairy and meat products to attract the consumer and to satisfy the strong demand for healthy foods. Mohair and cashmere will play an increasingly important role in the development of the goat industry.

The introduction of quotas for cow milk in the European Community in April 1984 immediately stimulated interest in alternative live stock enterprises. With a few exceptions, the goat can be farmed as a small cow. Indeed, many of the techniques employed routinely in cow milk production would, if applied more widely in goat herds, result in an improvement in the quality of goat milk and of goat milk products.

Nevertheless, with a strong demand for goat products there are exciting opportunities for the efficient producer, and the future for commercial goat production looks good.

We are grateful to many people for their encouragement during the preparation of this book. Professor Frank Raymond, who, as Chief Scientist for Agriculture and Horticulture, first stimulated our interest in goats when we were his colleagues in the Ministry of Agriculture, Fisheries and Food. Mr Alan Mowlem and Mrs Jas Barley of the Animal and Grassland Research Institute, for their foresight which led to the formation in 1984 of the Goat Producers Association, from which we gleaned much useful information. Alan kindly checked the whole manuscript and corrected our errors of fact. John Matthews and Michael Ryder also helped by checking Chapters 6 and 11 respectively. Their assistance is much appreciated. Richard Miles and Julian Grover of BSP Professional Books realised the need for a book on the commercial aspects of goat production and patiently waited for us to do the work. We thank the directors of Chiltern Beef Ltd, who allowed us to experiment in the rearing and marketing of goats for meat. Finally, a special thanks to Cherol Wilkinson, who not only tolerated our discussions whilst we wrote, but also typed the script with speed and accuracy.

J.M. Wilkinson
Barbara A. Stark
Marlow
Buckinghamshire

1   The Goat as a Producer of Food

Population

Goats play a significant role in producing food for the human population of the world. They do so particularly in those areas of the world where the climate is hot and dry. Thus the population of goats is concentrated in tropical and sub-tropical areas – in the developing countries of Africa and Asia – and in the warmer temperate regions around the Mediterranean and Central America (Table 1.1).

The world population of goats has increased steadily over the last thirty years, from under 300 million in 1950 to over 470 million in 1982. Most of this increase has occurred in those areas of the world where the

Table 1.1   Goat populations, 1984

	Number (millions)	% of total goat population
Tropical and sub-tropical areas		
Asia	255	56
Africa	151	33
South America	20	4
Temperate areas		
Europe	13	3
North and Central America	14	3
USSR	7	1
Oceania	0.5	0.1
World	460	

(Small discrepancies due to rounding)
Over 90 per cent of the goats in the world are in tropical and sub-tropical areas

Source: Food and Agriculture Organisation (FAO), *Production Yearbook 1984* Rome: FAO

goat is well established as an important domestic animal, and is most likely the result of improved management, leading to better nutrition and herd health.

Goats in dry regions of the world

Why is the goat predominantly found in the drier areas of the world? These areas contain a wide variety of plants, shrubs and trees, which provide a diverse supply of feed for animals at different times of the year.

Grasses grow in abundance in the tropics following periods of rain, whilst the leaves of shrubs and trees may remain green during periods of exceptionally dry weather.

Goats survive these huge changes in feed supply because they are adaptable – they graze when grasses are lush and abundant, but when the supply of grass is sparse, they browse the leaves of shrubs and trees. In this way they behave like other native ruminant species of the sub-tropics and tropics. Sheep, by contrast, are almost exclusively grazers. When the supply of herbage is reduced by prolonged dry weather, sheep are less likely to spend time browsing, and are at greater risk of starvation as a result.

The eagerness of goats to stand on their rear legs to reach branches of trees and shrubs is well known. They have even been known to climb trees to reach fresh green leaves!

Compared to cattle or sheep the requirement for water by the goat is relatively low. As a result, rumen volume is reduced and the dry matter content of rumen digesta is increased. Whilst a reduced requirement for water is an advantage in dry areas, it means that the goat may be at a disadvantage in terms of efficiency of rumen digestion of fibre. To compensate, the goat tends to be relatively more selective in its eating habits than cattle or sheep and is more likely to reject fibrous feeds of very low digestibility.

Goats in the United Kingdom and in Europe

There is evidence that the population of goats in the United Kingdom is increasing (Table 1.2). The annual registrations of goats with the British Goat Society rose from 2000 in 1970 to 10000 in 1980. The population of registered goats in the UK in 1985 was in excess of

80 000. Estimates in 1983 by the Ministry of Agriculture, Fisheries and Food (MAFF) indicated a total population of 27 446 goats on registered agricultural holdings in England and Wales (Table 1.2).

Table 1.2 Goat population on agricultural holdings in England and Wales

	1970	1975	1983
Goats in milk or in kid	3811	4047	NA
Other goats	5047	5854	NA
Total	**8858**	**9901**	**27 446**

NA = not available
Source: Ministry of Agriculture, Fisheries and Food (MAFF), Statistics Division, personal
communication 1985

In Europe the population of goats is increasingly becoming concentrated in larger herds. In France, for example, the average size of the herds recorded by the Institut Technique d'Ovine et Caprine (ITOVIC) rose from 53 to 64 between 1977 and 1982. Following a period of steady decline between 1930 and 1973, the goat population of France has now increased to over 1.5 million goats – about the same size as in 1900. In Switzerland a similar trend towards larger herds has helped to increase the population of goats to almost 1 million. In the Mediterranean region of Europe (Greece, Italy, Portugal, Spain) the population of goats increased between 1969 and 1979 from 8 million to 8.5 million. Most of this increase occurred in Greece and reflected a large expansion in the number of larger, intensively managed herds.

Table 1.3 Breeds of goat in the UK

Breed	% of total registrations, 1970 to 1980
British Saanen/Saanen	17
Anglo-Nubian	12
British Toggenburg/Toggenburg	12
British Alpine	5
Guernsey	1
Crossbred	53

Source: British Goat Society, personal communication 1985

Plate 1 Part of the British Saanen herd at the Animal and Grassland Research Institute, Shinfield, Reading. The Saanen and British Saanen types are the most common dairy goats in the UK. (*Courtesy of Food Research Institute, Shinfield.*)

The predominant breed of goat in Europe is the Saanen. This breed is popular because of its high milk yield. In upland areas the smaller Alpine and Toggenburg breeds are popular, whilst the lop-eared Nubian is more common in southern Europe. Crossbred goats are very common; they comprise over half the registered population of the UK. Many herds have a complete mixture of pure and crossbred goats, with selection and breeding based on individual merit rather than aiming towards a uniform type of animal.

Output

World goat milk and goat meat production, at 7.7 and 2.1 million tonnes respectively, represents 1.6 and 3.8 per cent of total milk and meat production from farmed ruminants, i.e. cattle, buffalo, sheep and goats (Tables 1.4 and 1.5). Most of the goat milk and goat meat is produced in Asia, with China being the principal Asian producer. Europe, with only 3 per cent of the world population, produces over 20

Table 1.4 Goat Products, 1984

	Milk	Meat (000 tonnes)	Fibre	Hide
Tropical/sub-tropical areas				
Asia	3546	1222	NR	236
Africa	1483	609	NR	109
South America	136	64	NR	14
Temperate areas				
Europe	1658	86	NR	13
North and Central America	351	30	NR	8
USSR	330	30	NR	9*
Oceania	NR	2	NR	0.2
World	7505	2042	20[1]	389

(small discrepancies due to rounding)

NR = Not recorded
* = Estimate
Sources: Food and Agriculture Organisation (FAO), *Production Yearbook 1984* Rome: FAO
(1) Ryder, M.L. (1986) 'The goat' *Biologist,* 33, 131–139

Table 1.5 Goats and their products as a percentage of all ruminants, 1984

	Number	Products			
		Milk	Meat	Fibre	Hide
Europe	4.3	0.9	0.7	NR	1.0
World	15.3	1.5	3.7	1	4.9

NR = Nor recorded
Sources: Food and Agriculture Organisation (FAO), *Production Yearbook 1984* Rome: FAO
Ryder, M.L. (1986) 'The goat' *Biologist* 33, 131–139
International Wool Secretariat, personal communication 1986

per cent of total goat milk but only 4 per cent of total goat meat. The extensively managed herds of Asia and Africa produce almost 90 per cent of world goat meat, whilst the more intensively managed herds of

Europe are geared principally to the production of milk, with meat as a by-product.

Average yields of milk per head in Europe vary from as little as 100 litres in the Mediterranean region to over 550 litres in northern and central Europe. For example, herds recorded by ITOVIC in France averaged 583 litres per head in 1982.

Statistics for world production of goat fibre are difficult to obtain, but it is estimated that 3000 to 4000 tonnes of cashmere and about 15 000 tonnes of mohair are produced per year. A high proportion of the total mohair and cashmere produced is imported into the UK for processing and in 1985, 7356 tonnes of mohair and 2188 tonnes of cashmere were imported.

The present world supply of mohair is derived mainly from South Africa, Turkey, and Texas in the USA. Cashmere is produced in the Himalayas and other mountainous regions from the secondary (very fine) hair follicles of the goats in those areas.

Goat hides are used for the production of leather goods, and goatskin rugs. They are also used locally for clothing and in furniture.

Organisations

In many countries there are national and local organisations which serve the commercial goat producer. For example, in the USA the American Dairy Goat Association and the American Goat Society provide assistance through milk recording, breed improvement schemes, enterprise costing programmes and technical information.

Organisations offering artificial insemination services for goats are in operation in several countries. In Norway a progeny-testing programme is in operation to identify elite sires.

The establishment of a network of recorded herds in France by ITOVIC has undoubtedly helped producers to identify key components of economic success and to work towards clearly defined economic goals. But government technical support for the goat industry in other European countries is not yet as well advanced as in France, despite the existence in the European Community of support programmes, in the form of either credit assistance or capital grants for farm development and, since 1981, of a sheep and goat meat price support regime.

In the UK the British Goat Society has played a leading role in identifying breed type, registering bloodstock and assisting producers in the exportation of livestock.

The Goat Producers Association serves to provide the commercial producer with technical support for both production and marketing of goat products. Caprine and Ovine Breeding Services (COBS) offers an artificial insemination service. The Agricultural Development and Advisory Service (ADAS) of the Ministry of Agriculture, Fisheries and Food have specialist advisors in dairy goat husbandry, and in goat nutrition. ADAS publish technical information on these aspects of goat production, and can advise individual commercial producers.

Research into goat production is under way in all countries, though the amount of work varies according to the relative importance of the goat industry. Government research institutes and University departments of animal science in the USA and in Europe are conducting research into nutrition, physiology, reproduction and growth using goats, but often the goat is used because it is a convenient small ruminant. Thus the work tends to be of a strategic or basic type rather than of direct application to the goat industry. Nevertheless useful applied work is being done by several centres in the USA, by the Institut National de Recherche Agronomique (INRA) in France, and in the UK at the Animal and Grassland Research Institute, Reading, and the Hill Farming Research Organisation, Edinburgh.

Much of the European research and development work is co-ordinated by the Sheep and Goat Commission of the European Association of Animal Production. New information is regularly reported at their meetings and is of value to other research groups, advisers, teachers, and commercial producers.

Future trends

The trend towards larger herds will continue in common with other animal production enterprises. Developments in nutrition and reproduction will mean that yields of milk and meat will increase, that seasonal peaks and troughs in output per herd will be less pronounced, and that product quality will improve.

As the goat industry in the temperate countries of the world becomes technically more advanced, new information will be available to the developing areas of the world. In this way the goat may become both a more efficient and a more important supplier of high quality food to a growing human population.

Further reading

Gall, C. (1981) 'Socio-economic role of goat husbandry in the Mediterranean region' *Paper S 1–3 32nd Annual Meeting of the European Association for Animal Production* Yugoslavia.

Sigwald, J.P. and Lequenne, D. (1983) 'Resultats contrôle laitier 1982' *La Chèvre* 138: 24–25.

Throckmorton, J.C. (1981) 'The potential for goat production' (in: *Recent Advances in Animal Nutrition in Australia* (Ed) D.J. Farrell) Universitity of New England Publishing Unit, Armidale, NSW, Australia.

Van Soest, P.J. (1982) *The Nutritional Ecology of the Ruminant* O & B Books Inc., Corvallis, Oregon, USA.

2 Systems of Commercial Goat Production

Planned systems of production

Planned production is the key to profitability. In addition to knowing how to produce a high quality product and being able to sell it at a reasonable profit, it is necessary to identify the technical factors which contribute to economic success or failure. For example, a producer of goat milk contaminated with *Salmonella* would soon go out of business if the milk were sold unpasteurised and given to children. Correct nutrition, good herd health and a high standard of hygiene mean that the product will be of a higher quality and produced at a lower unit cost than that from poorly-fed, diseased animals.

Production systems need to be developed to suit the resources of the farm. There is less sense in building an intensive milk production unit on a hill farm with limited land suitable for grass conservation than in doing so on a lowland farm where winter feed can be produced at low cost, and where the winter period is of shorter duration.

The size of the unit should bear some relation to the financial and physical resources of the farm. It is better to grow the business steadily than to borrow excessively, over-stretch available labour, or over-stock the land at the outset. Perhaps most important of all – the system of production should be designed to suit the specifications of the product. Profit is made by solving a problem or meeting a need. The market for the product must be identified at the outset and thoroughly researched so that targets for turnover can be set and, hopefully, reached. There is little point in producing goat milk or meat if nobody wants to purchase the product. Customers will be lost if the product is of variable quality, or if supplies are erratic.

Planned selling implies knowing how much product the customer is likely to require, and organising the production schedule accordingly. With an animal which is a seasonal breeder in temperate countries, and with no official price support, fluctuations in output – and hence in price received for the product – can easily erode profits unless care is taken to ensure that the system is well designed for the particular product and its market.

Milk

Traditionally, goats have been kept in relatively extensive systems of production on smaller farms. Around the Mediterranean area, goatherds still tend their animals and ensure that they graze defined areas of land. In northern Europe herds of mountain goats are kept indoors in winter and grazed on the hills in summer. Often the animals remain on the hill throughout the growing season, are milked by hand, and the milk is made into cheese before being transported back to the valley for sale.

In lowland areas goats are usually kept indoors for most, if not all, the year. Larger herds are milked through parlours. Grazing, if it is practised, is confined to pasture near to the milking parlour. The animals are given concentrates based on cereal grains and high-protein feeds, together with hay *ad libitum*. Larger lowland herds are increasingly turning to silage as the preferred form of conserved forage. In the southern states of the USA, and in southern areas of Europe, goats commonly receive by-products such as apple waste, citrus pulp and cottonseed hulls.

It is significant that goat herds have developed for milk production in areas where small-scale cheese and yoghurt production is also relatively well developed. Not surprisingly, skills developed with cow cheese and yoghurt production can be adapted easily to goat milk products. Further, in countries where there is a tradition of making dairy products with distinct local characteristics, such as France, it is relatively easier to interest potential customers in new products than in regions where such traditions are less well established.

Typically, herds of milking goats are small – either because the farm itself is small or because the animals are hand-milked. However, the transition to machine milking and all-year-round housing allows considerable expansion without necessarily increasing the input of labour. Housed herds can be bred out-of-season, either by manipulation of lighting pattern or by hormonal stimulation of ovulation, to give a more even year-round pattern of milk output.

Meat

Systems of goat meat production are poorly defined in many temperate countries. However, in countries such as France, Greece and Italy, where goat meat is produced in quantity, there is a speciality market for kidlet, particularly at Easter. Kidlet is produced on farms where milking goats are kept. Male kids and surplus females are reared either as suckled kids or are separated from their mothers and given whole milk or a milk substitute. The kids are slaughtered at between 8 and 16 weeks of age, at 10 to 20 kg liveweight. In Norway, smoked goat meat is produced, some of which is exported to other European countries.

A problem associated with the production of goat meat as a by-product of the dairy herd is that the conformation of the Saanen-type is not considered attractive for meat. This situation is analogous to that in the dairy cow, where the introduction of Holstein blood has led to discrimination against calves from Holstein cows because of their alleged unsuitability for beef production. A possible solution might be to develop 'meaty' types of sire breeds for crossing with the Saanen.

The goat meat industry is poorly developed in the UK, despite the presence of substantial numbers of people who are familiar with goat meat and would eat it by choice if it were available regularly. Not only are there opportunities for producers to fulfil demand for home consumption, both domestic and in catering establishments, but there are also opportunities to meet a growing demand for kidlet from other countries – principally those around the Mediterranean and in the Middle East.

If specialist kidlet producers can identify suitable markets, then appropriate systems of production are likely to develop. For example, kids surplus to requirements for milk production could be reared on milk substitute or multiple-suckled on older goats which would otherwise be culled from the herd. In hill and upland areas suckler herds of goats could perhaps be established for kidlet production in much the same way as sheep flocks currently specialise in producing suckled lambs in late spring and summer.

Fibre

There are two quite distinct types of goat fibre – mohair, which is similar to fine sheep wool, and cashmere, the very fine hair from the secondary follicles of cashmere-type breeds. Herds of Angora goats are kept for mohair production, with meat as a by-product from culled females and males. The animals are grazed extensively and are normally sheared twice each year. With increased interest in high-value fibres from goats, the possibility arises of producing fibre as a by-product from dairy herds by crossing Saanen-types with Angoras.

The Angora × Saanen offspring have a more suitable conformation for meat than the pure Saanen, and thus there would appear to be considerable potential for introducing Angora blood into a proportion of the milking herd, to give not only fibre but also improved meat animals.

Cashmere is produced in China, Turkey and Russia as a by-product from goats kept for both milk and meat. Feral goats have also been found to yield cashmere. However, the commercial feasibility of cashmere production in hill and upland goat herds in Europe has yet to be fully established.

Integrated systems

Successful production of goat milk and milk products involves standards of husbandry and hygiene which are comparable to those currently operated in dairy cow herds. With the introduction of quotas for cow milk in the European Community (EC) there may be attraction in the possibility of developing integrated lowland farming systems in which milk is produced from both cows and goats on the same farm. Currently there are no EC quotas for goat milk or goat milk products. Indeed, demand for these foods would appear to be increasing whilst that for cow milk and cow milk products, with the exception of yoghurt, is not.

Diversification in milk production into goats may therefore offer a way of increasing the size of the business in the face of physical controls on cow milk production. There may be opportunities for spreading fixed costs such as labour, buildings and machinery over both cow and goat enterprises.

In hill and upland areas the complementary grazing habits of goats and sheep may be of considerable benefit in increasing output. Recent research in New Zealand and the UK indicates that whilst sheep prefer to graze clover, goats do not, preferring to select in favour of long grasses and rushes when they are present in the sward. Thus mixed grazing of goats together with sheep may benefit the sheep enterprise by reducing grazing pressure on the clover component of the pastures. In addition, by the goats' preferential grazing on weeds such as rushes, the proportion of grasses of higher feed value to sheep may be increased by the presence of goats.

Further reading

Mowlem, A. (1984) 'Milk and meat production from goats' *Paper 86, British Veterinary Association Congress, Stirling, Scotland* September 1984.
Mowlem, A. (1984) 'Goats for meat and fibre' *Smallholder* **2**: 10.

3 Products

The commercial producer is also the salesperson of the products from the herd – either to a wholesale distributor or to the retail customer. Whichever method of sale is adopted, the more the producer knows about the composition of the product the better. The informed producer can educate potential customers and can handle enquiries concerning further processing, such as cooking in the case of milk, milk products or meat, or spinning in the case of fibre.

In this chapter the composition of goat products is described and discussed in comparison to similar products from other ruminants. Systems of goat milk, meat and fibre production are discussed in more detail in Chapters 9, 10 and 11 respectively.

Milk and milk products

Goat milk has a similar gross composition to that of cow milk (Table 3.1), although the milk from the Anglo-Nubian breed typically contains up to two percentage units higher concentration of fat. There are differences between the ruminant species in the types of protein in milk – goat milk contains more β casein and less α casein than cow milk. However, the overall amino acid composition of the protein fraction is similar between goat milk and cow milk.

Table 3.1 Typical composition of goat, cow and sheep milk (%)

	Goat	Cow	Sheep
Total solids	11.9	12.8	19.4
Fat	3.9	3.9	8.3
Protein	2.9	3.3	5.4
Lactose	4.3	4.8	4.8
Ash	0.8	0.8	0.8

The proportion of short and medium chain length fatty acids is higher in goat milk than in cow milk and there is also a higher proportion of small fat globules (Table 3.2).

Table 3.2 Distribution of fat globules in goat, cow and sheep milk

Size of fat globules (µm)	Proportion of total globules (%)		
	Goat	Cow	Sheep
1.5	28	11	29
3.0	35	33	40
4.5	20	22	17
6.0	12	18	12
>6.0	5	16	2
Average diameter	3.5	4.5	3.3

Source: Parkash, S. and Jenness, R. (1968) 'The composition and characteristics of goats milk: a review' *Dairy Science Abstracts* **30**, 67–87

Plate 2 Goat milk products are increasing in popularity throughout Europe and North America. Liquid milk, cheeses and yoghurts are now available in many supermarkets, delicatessens and health food shops. (*Courtesy of Food Research Institute, Shinfield.*)

Goat milk is substantially lower in both total fat and total protein than ewe milk.

Cheese yields from goat milk are similar to those from cow milk, typically about 10 kg hard cheese and 15kg soft cheese per 100 kg of milk. The composition of goat cheese is also comparable to that of cow cheese, but its physical texture is different. Goat milk forms a softer, more friable curd when acidified and as a result cheeses from goat milk tend to be softer than those made from cow milk.

Yoghurt is a traditional fermentation product of goat milk. Essentially its composition is similar to that of the milk from which it has been made, except that lactose is fermented to lactic acid.

Butter and cream from goat milk are noticeably different in appearance from cow products, in that both are white. In other respects they are similar in gross composition to cow products.

Typical yields, solids content and fat content of goat milk products are given in Table 3.3.

Table 3.3 Typical yield and composition of goat milk products

	Yield (kg/100 kg milk)	Total solids (% of fresh weight)	Fat
Cheese:			
hard curd	9 to 12	62 to 66	28 to 34
soft curd	11 to 15	40 to 50	18 to 25
Yoghurt (whole milk)	100	12	4
Cream[1]	9[2] to 19[3]	40[4]	36[4]
Butter	40	83	80

(1) Produced using a separator
(2) Double cream
(3) Single cream
(4) Whipping cream

Meat

The most striking characteristic of the goat carcass is the virtual absence of subcutaneous fat. This does not mean that goats do not store excess energy as fat but rather that its distribution is different from that of sheep (Tables 3.4 and 3.5).

Table 3.4 Carcass composition: goat compared to sheep

	Sheep	Goat
Carcass weight (kg)	15	12
Composition (%)		
Muscle	58	62
Bone	14	19
Fat	28	19[1]

(1) The lower proportion of carcass fat reflects a higher proportion of abdominal (non-carcass) fat – see Table 3.5
Source: Wood, J.D. (1984) 'Composition and eating quality of goats meat' in: *Developments in Goat Production 1984, Proceedings of the Inaugural Conference of the Goat Producers' Association of Great Britain*, 17 April 1984

Table 3.5 Distribution of body fat: goat compared to sheep (%)

	Sheep	Goat
Subcutaneous fat	43	16
Intermuscular fat	33	37
Abdominal fat[1]	24	46

(1) Non-carcass fat
Source: Wood, J.D. (1984) 'Composition and eating quality of goats meat' in: *Developments in Goat Production 1984, Proceedings of the Inaugural Conference of the Goat Producers' Association of Great Britain*, 17 April 1984

At similar carcass weights goats may be expected to produce carcasses with a lower proportion of carcass fat and a higher proportion of non-carcass fat in the abdominal cavity (Table 3.5). The lower proportion of fat in the goat carcass is reflected in a higher proportion of bone than in sheep (Table 3.4).

Intensively reared castrated male goat kids may be expected to deposit considerable amounts of fatty tissue in the later stages of their growth. Thus in trials in which kids were slaughtered at either 24 kg or 36 kg, the proportion of fat in the carcass increased from 17 per cent at the lower slaughter weight to 24 per cent at the higher slaughter weight. Gross carcass composition at the higher weight was similar to that of sheep carcasses of similar weight.

The relatively high bone content of the goat carcass reflects the longer legs and bodies of goats compared to sheep. This is considered undesirable by the meat trade in the UK and as a result goat carcasses are likely to be undervalued. A possible alternative approach is to bone out the goat carcass so that the meat can receive further processing before being consumed.

Fibre

Goats produce high quality fibre products which are in considerable demand for use in the textile industry. Cashmere, with its extreme fineness, is especially valued.

Typical yield and composition of mohair and cashmere are compared to the wool from the adult merino sheep in Table 3.6. Because cashmere is derived from secondary follicles alone, the yield is low, typically about 400 g of unscoured fleece per adult. The proportion of clean fibre is also somewhat lower than that which would typically be obtained from mohair from adult Angora goats.

Table 3.6 Typical composition and characteristics of fibre: goat compared to sheep

	Sheep	Goat	
	Wool[1]	Mohair	Cashmere
Yield (kg unscoured fibre)	4	4	0.2 to 1.5
Clean fibre (%)	70	80	60
Kemp (% total fibre)	negligible	<10	negligible
Length (mm)	100	140	40
Fineness (μm)	20	36	16

(1) Adult merino

The annual yield of mohair is similar to that of merino wool. Fibre size and strength are typically greater for mohair than for merino wool. Goat hair secondary follicles must be less than 19 microns in diameter to be classified as cashmere.

Plate 3 Young Angora goats grazing extensively. These goats produce high-quality mohair fibre. (*Courtesy of International Mohair Association.*)

Other products

Goat skin can yield very high quality leather, particularly from young kids. Skins are also salted for use as rugs, and for use in the manufacture of musical drums.

4 Breeding

The domestic goat has been farmed since 7000 BC, principally in the warmer areas of the world (see Chapter 1). Its predecessors were probably concentrated in wild populations in central Asia and around the Mediterranean. The natural inquisitiveness of the species probably accounts for its being one of the first animals to be domesticated.

Breeds

A wide variety of breeds exists today, ranging from extreme dairy types with narrow, angular bodies, short white coats and erect ears, to long-coated breeds which look like sheep and are kept for their highly valued fibre (Table 4.1).

The most popular breed for milk production is the Saanen and improved Saanen-types such as the British Saanen and the French Saanen. The breed holds the world record for milk production, at over 3500 litres in 365 days. Typically Saanen-types yield 800 to 1300 litres per lactation.

Most commercial herds also include some Anglo-Nubian goats because these animals produce milk of a higher fat content than that of Saanens. The fat level in the milk of Anglo-Nubians is normally above 4.5 per cent. Thus the breed complements the Saanen and can be particularly useful if the major product is cheese.

The breeds kept mainly for meat are smaller than those kept for milk. They occur in Asia, Africa and South America. Elsewhere, meat is a by-product of the dairy herd. Male kids and surplus female kids are usually reared for meat. Culled adults are also a source of meat.

Perhaps the most highly prized goats are those of the Angora and Cashmere breeds which are kept for their valuable fibres. The word Cashmere denotes both the Cashmere breed itself, and also the very fine fibre which can be harvested from several other breeds. Both Angora and Cashmere goats originated in Asia, but the high value of

Table 4.1 Breeds of goat for milk, meat and fibre

	Country of origin	Colour	Average weight (kg)	Principal features
Breeds kept mainly for milk				
Saanen/ British Saanen	Switzerland	White	65 to 75	High milk yield; short and fine coat; angular body; erect ears; quiet temperament
Anglo-Nubian	UK	Red tan, mottled or dappled tan	60 to 70	High milk fat content; short coat; narrow body; long, pendulous ears; convex nose
Alpine/British Alpine	Alps	Brown/ black	45 to 55	Facial stripes; short coat; erect ears
Toggenburg/ British Toggenburg	Switzerland	Fawn/ brown	45 to 55	Facial stripes; short coat; erect ears
Golden Guernsey/ English Guernsey	UK (Channel Islands)	Reddish brown	40 to 50	Long coat; erect ears; straight nose
La Mancha	USA	Mixed	60 to 75	Polled, vestigial ears; short coat
Nordic	Norway, Sweden, Finland	White	45 to 55	Long coat; erect ears
Breeds kept mainly for meat				
Criolla	South America	Mixed	30 to 45	
Ma T'ou	China	White	40 to 60	
Pygmy	Africa	Brown and white	20 to 25	
Dwarf East African	E Africa	Mixed	25 to 30	
South China	SW China	Black	20 to 40	Twisted horns
Dara Din Panah	India	Black	40 to 45	Long coat
Breeds kept mainly for fibre				
Angora	Central Asia	White	45 to 55	Ears horizontal or drooping; long coat, hanging in lustrous locks or ringlets; long, curling horns
Cashmere	Central Asia	White	35 to 45	Erect ears; long, twisted horns

the fleece has meant that Angora goats are now farmed in the USA, Australia, New Zealand, Southern Africa and, more recently, in Spain and the UK.

Breeding objectives

The principal objective is to breed for high yield. In dairy goats this means milk output if the product is liquid milk or yoghurt. If the product is cheese then the comparable criterion might be yield of total milk solids. For meat production a major aim is to have as high a rate of growth as possible. Similarly, rate of fibre growth is clearly of prime importance in achieving high fibre yield.

A secondary, vital objective is to breed for high reproductive efficiency. The larger the population of young born each year, the higher the selection pressure and the greater the genetic gain per year, provided there is adequate variation between individuals.

Selection pressure is only possible when it is the intention to maintain herd size and to replace a proportion of the herd each year by young females which have been specially selected on the basis of performance information. The higher the replacement rate the higher the selection pressure, *provided* the next generation are of superior breeding value. To achieve this it is necessary to know not only the performance of the dam but also that of the sire. That is, the performance of the sire should have been recorded prior to his selection as a suitable male. In the case of growth rate and fibre yield this is possible by *performance testing* a group of young males of similar initial age over the same period of time. Preferably the testing should be carried out at a testing station where nutrition and environment are constant.

As milk production is a sex-linked trait, males have to be evaluated on the performance of their dams and of their female progeny. As yet no satisfactory indirect selection method exists for assessing the milk yield potential of male goats. The best solution is a combination of selection on dam performance records and *progeny testing*. If possible a short-list of the best males should be provisionally drawn up based on the breeding values of their parents. These are then progeny tested and the best of these are used to produce the next generation of sires and breeding females.

Breed improvement

Selection within a breed for specific performance traits should result in improvement in those traits over a number of generations. The rate at which improvement occurs is known as the rate of *realised genetic gain*. Genetic gain may not always be obvious because of overriding effects of environmental, managemental or nutritional factors. Also, the rate of genetic gain in one trait may be reduced because of simultaneous selection for other traits. Thus the best milk yielders may not have the highest reproductive efficiency, and if selection for reproductive efficiency is also carried out, animals of lower milk yield will be retained, thus reducing the average improvement in milk production. Therefore, the rate of improvement in milk yield will be highest when yield is the only trait under consideration, when the selection pressure is high and when both females and males are tested adequately.

Performance recording

Clearly it is important to record the yield of important performance traits in a well organised breeding programme. If possible, collaborators should agree to standardise on as few traits as are necessary to achieve improved economic efficiency.

Group testing programmes are in operation in several countries. They may involve independent testing of milk yield, milk composition, growth rate and reproductive efficiency to verify the producer's own recording procedures. Performance recording can be computerised centrally so that individuals and herds may be compared on the basis of rolling averages. Sires can then be speedily identified on the basis of the performance of their progeny. Centralised recording schemes are particularly useful if sires are evaluated through artificial insemination, or where individual herd size is small. Different breeds and sire progeny groups can be easily compared, and superior stock identified on a national basis.

The principal measure of performance in dairy goats is the lactation record, which is usually corrected to a standard number of days. This standard lactation may be the same length as for dairy cows, i.e. 305 days, or may correspond to the calendar year of 365 days. There are obvious advantages in equating goats with cows so that commercial goat producers can take advantage of computerised recording schemes already in operation for dairy cattle.

Since milk production tends to increase with age until about the third lactation, this effect can be taken into account by adjusting the performance record to a standard age. Similarly, records may be adjusted to a standard kidding season to remove seasonal effects on lactation.

Other influences on lactation yield can be removed by expressing the yield of each individual as the difference between her record and that of the other females in the same herd in the same year. This is known as the *herdmate deviation*. This technique is particularly useful as a guide to selection within herds containing a number of different breeds and crosses.

There is scope for using centralised performance recording schemes already in operation with other livestock species to develop more effective breed and sire evaluation for goats. In addition, the advisory and research expertise which already exists for the livestock industry would be available to goat producers if they were willing to participate in existing recording schemes. In this way the extension organisations would be encouraged to become more involved in the goat industry, with a consequent improvement in the quality of their services.

Performance testing

The technique of performance testing is well established for beef bulls. Traditionally, groups were assembled at testing centres and reared under uniform conditions. Weight gain was recorded from 200 to 400 days of age and bulls were ranked on daily gain and weight for age.

More recently, performance testing on farms has become more popular as individual breeders have sought to demonstrate continuous improvement in the herd over successive generations. With careful management and attention to standardising the conditions of testing, on-farm performance testing can offer considerable benefits to breeders of goats, particularly those breeds suitable for fibre and meat production.

Progeny testing

Recording the milk yield of the dam can only indicate half the total genetic influence on performance, since the remaining half has been contributed by the sire. More rapid propress can be made if the 'milk yield' of the sire is assessed prior to his use extensively as a breeding animal. This involves testing a population of his daughters and comparing their performance with that of the average for the breed.

Since most commercial herds of goats are relatively small, progeny testing either involves a group of breeders co-operating to test a small number of sires selected on the basis of pedigree information, or artificial insemination.

A compromise has to be struck between the number of progeny per sire and the number of sires to be tested. If the number of progeny per sire is too small, the accuracy of the test will be reduced. If on the other hand the number of progeny per sire is too large, only a small number of sires can be selected and compared.

Work in the USA has shown that the optimum number of progeny tested per sire is 15. Smaller groups cause problems in establishing sufficiently well-balanced groups to make valid comparisons. Larger groups reduce the number of males on test and consequently the selection pressure.

Co-operative breeding schemes involving natural service require good organisation so that males are moved around during the mating season to ensure an adequate number of daughters per male. In some schemes it is necessary to extend the test of a group of males over two mating seasons to achieve a sufficient number of progeny per male. This has the effect of extending the generation interval between males, and reduces the rate of genetic gain.

Having selected proven sires, it is desirable to use them for a single season to minimise the sire–son generation and also the risk of inbreeding.

Part-lactation records

If full 305- or 365-day lactation records are used as the basis for selecting sires then they cannot be used in the mating season which occurs during that particular lactation. This means the loss of a full generation before the information can be used and limits the rate of genetic progress.

It is very useful therefore to use part-lactation records as the basis on which males are selected. Ideally a progeny test should be evaluated using 4-month lactation records extended by regression methods to reflect the full lactation period. Whilst there may be some reduction in accuracy, this is more than compensated for by the shortening of the generation interval on the male side by one year compared to the use of complete lactations.

The effect of completing a progeny test in one rather than two years, and of using part-lactations, is illustrated in Table 4.2, and serves to emphasise that co-operation between breeders, good organisation, and rapid processing of data are the essential elements of successful sire evaluation.

Table 4.2 Relative genetic gain per year due to selection of sires over 1 or 2 years, with or without selection based on part lactation records

Number of mating seasons	Use of part lactations	Generation interval (years sire–son)	Relative annual genetic gain[1]
1	No	5	100
1	Yes	4	125
2	No	4	125
2	Yes	3	167

(1) One service year per proven sire
Source: Steine, T.A. (1982) 'Principles of selection for milk production in dairy goats' in: *Proceedings of 3rd International Conference on Goat Production and Disease*, Tucson, Arizona, USA pp 19–25

Heritability

Genetic progress is not only the result of selection, it is also a reflection of the extent to which desired attributes are inherited. The *heritability* of a particular trait is the proportion of the variation due to genetics. Traits such as coat colour are determined mainly by genetics and so have high heritabilities. Other traits such as fertility are subject to environmental influences and so have low heritabilities.

Selection for traits with low heritabilities is unlikely to result in rapid genetic gain because animals with superior genetic ability are difficult to identify. Often what appear to be superior parents do not produce superior offspring because inheritance is of minor importance.

Because factors other than inheritance often dominate, estimates of heritability vary from test to test. Milk yield and milk fat percentage probably have heritabilities of about 0.3, i.e. 30 per cent of the variation in these traits is due to genetics. These values are similar to those for dairy cattle.

Weight gain in kids may be expected to have a somewhat higher heritability – perhaps as high as 0.5, though information on the heritability of this trait is scant.

The heritability of fibre yield in Angora goats is relatively high, at 0.4. But low values have been found for traits such as fibre diameter. In one USA flock, high heritability values were found for kemp score, staple length and fibre yield (the range was from 0.4 to 0.8), indicating that in this particular instance good progress could be expected for selecting for these traits. However, in other herds environmental influences were more evident and heritability values for the same traits were much lower.

Repeatability

The degree to which a characteristic is repeated over time is known as its repeatability. This is an indication of the ability of an animal to maintain its superiority in a particular trait from year to year.

Estimates of repeatability are normally higher than those for heritability. For example, although the heritability of fibre diameter in Angoras is low (0.19), its repeatability is high (0.72). Thus, having identified an animal with fine, high value fibre in a particular year, the animal is likely to maintain its superiority in future years.

Genetic correlations

Sometimes it is useful to measure a performance trait indirectly – either because it is sex-linked or because the desired characteristic is relatively difficult to measure. Not surprisingly, relatively high correlations have been found between staple length and fibre yield, which means that if yield itself cannot be measured, samples of fibre can be measured for length as an indication of likely yield. This may be particularly useful if there is a need to select animals for breeding before shearing takes place.

In contrast, studies have shown a low correlation between reproduction (difficult to measure reliably) and milk production (easier to measure). This is unfortunate in that selection for milk yield is unlikely to improve reproductive performance, but fortunate in that it is unlikely to reduce it.

Rate of genetic gain

The actual rate of genetic gain in a desired trait depends on its heritability, repeatability and the selection pressure exerted on the population. Theoretical rates of gain can be calculated from knowledge of these factors, and in practice the actual rate of gain is about 80 per cent of the theoretical rate.

Depending on the design of the breeding programme and the size of the breeding population, the theoretical rate of genetic gain in milk yield will be between 1.5 and 3 per cent per year. The higher rate relates to full recording of both males and females. Studies have indicated that rates of gain in excess of 3 per cent could be achieved if progeny-tested males are used to sire young males which in turn are then selected so that the best offspring sire most replacement females.

In practice, actual annual rates of gain have been found to be very variable. For example, values as low as 0.22 and 0.24 per cent for milk yield and milk fat respectively were found in surveys of goat populations in the USA. These low values reflected the lack of opportunity until recently to identify and use superior sires.

By contrast, work in Norway indicated rates of genetic gain for milk yield in excess of 1 per cent in herds co-operating in breeding schemes which involved both sire and dam selection.

The influence of the sire on genetic gain may be substantially increased if artificial insemination (AI) is employed rather than natural service. However, it is vital that the sires used for AI are of superior genetic merit, and they should have been selected following accurate progeny testing.

Embryo transfer involves the planned mating of a superior sire with a superior dam and is a useful technique to increase the maternal influence on a population. Embryos are collected following the mating and are transferred to recipients. The technique is particularly useful when there is a need to increase rapidly a population of high-value animals.

Both techniques are described in more detail in Chapter 5.

Crossbreeding

With a high proportion of crossbred goats already in use in many goat herds, it is interesting to speculate on the effect of crossbreeding on the rate of genetic gain. Limited information from studies in India suggests that crossbreeding is likely to be of value when the objective is to increase milk yield or growth rate, but of relatively little value when the objective is to increase the concentration of fat or total solids in milk. However, the milk production trials did not involve the use of breeds such as the Anglo-Nubian, with its characteristically higher milk fat content.

There is a clear need to establish the extent to which heterosis, or hybrid vigour, can be exploited in commercial goat herds. In addition, the opportunities for using crossbreeding to enhance the ability to produce both milk and fibre, or milk and meat, or meat and fibre, should be investigated.

Further reading

Production Testing for Dairy Goats (1980) *Western Region Publication No. 40* Cooperative Extension Service, United States Department of Agriculture.

5 Reproduction

Goats are prolific breeders. Typically they produce twins, and in well-managed herds a kidding percentage approaching 200 is not uncommon. In tropical areas of the world the breeding season is extended through the complete year. In the tropics three litters in two years is the normal pattern of reproduction.

Seasonal reproduction

In temperate regions the natural breeding season is confined to the period of decreasing day length (i.e. the autumn). The period is shorter the farther from the Equator. Thus in southern Europe and the southern states of the USA, females begin to show signs of oestrus in late July and August, whilst the season normally commences in mid-September in northern Europe and in the northern areas of North America.

The major consequence of seasonal ovulation is that there is a period of anoestrus from February to May when females do not cycle and are effectively infertile. This places limitations on providing a year-round supply of milk – though fortunately goats can be induced to breed out of season (see below).

Fertility

Efficient reproductive efficiency depends on having females that are fertile, i.e. capable of conceiving. Fertile females are those which are ovulating and showing signs of oestrus. Normally oestrus and the accompanying ovulation occurs once every 19 to 23 days, with the actual period of oestrus lasting about 24 hours. During this time the female usually shows obvious behavioural signs of agitation, bleating, tail wagging and being mounted by other females.

Ovulation occurs within 12 hours of the end of behavioural oestrus. Ideally the male should be introduced towards the end of the oestrus period, and again 12 hours later if the female will still accept him. Artificial insemination should occur either towards the end of the oestrus period or the next day.

After kidding there is usually a period of anoestrus which may be postpartum anoestrus, seasonal anoestrus or a combination of both factors. Frequent suckling can be reflected in anoestrus, but in goats bred naturally and milked twice daily the predominant effect on anoestrus is likely to be seasonal.

Initial stimulation of reproductive activity is usually achieved by introducing a male. Even if the male is kept out of sight, the odour of a male can induce oestrus in the female. For example, a group of females can be brought into near synchronous oestrus by exposing them to a piece of towelling which has been well rubbed on the head of smelly, sexually-active male. The towelling is placed in a large wide-mouthed jar and sealed. At the chosen time, the jar is opened in the pen containing the females so that they can smell it. They normally show intense interest in the jar's odour and most will be in standing oestrus within 2 to $2\frac{1}{2}$ days. Additional stimulation of reproductive activity during the natural season can be brought about by drying off lactating females that are not already pregnant.

Induction of oestrus can also be achieved at any time of the year by using progesterone-releasing intravaginal sponges designed for sheep. The sponges are carefully inserted into the vagina and left there for 21 days. The females are given an injection of 400 to 600 i.u. of pregnant mare serum gonadotrophin (PMSG) at the same time as the sponge is removed. The animals will ovulate 24 to 30 hours after removal of the sponge. This technique is particularly useful for artificial insemination and when there is a need to produce batches of kids born at the same time.

Flushing

Goats respond similarly to sheep in that regular cycling and ovulation is encouraged by good health and a rising plane of nutrition ('flushing') at the onset of the breeding season. Young, underweight or underfed females tend to produce fewer ova and are therefore more likely to give birth to large, single kids, with accompanying problems of dystocia at birth.

Body weight and condition at the start of the breeding season are probably the two most important factors affecting subsequent kidding performance. Young females, mated in their first year of life, should have reached about 60 per cent of their adult weight (approximately 40 to 45 kg) at the time of mating.

Nutrition prior to mating is clearly important in determining the extent to which flushing is achieved. In sheep, flushing traditionally lasts for about 3 weeks prior to the introduction of males. Flushing for 8 weeks rather than 3 weeks is likely to give heavier females at mating and stronger kids at birth.

Synchronising oestrus

Batch kidding is a very useful aid to management, particularly in larger herds. Intravaginal sponges (see page 31) or injection of either prostaglandin F2 α or luteinizing hormone-releasing hormone may be used to induce simultaneous ovulation. It is advisable to consult a veterinary surgeon in the first instance so that the correct procedures are adopted in relation to the natural breeding season.

Out-of-season reproduction

The need to maintain a uniform supply of milk throughout the year has led to the increased use of out-of-season breeding in commercial herds. The two main techniques for artificially inducing ovulation are hormone treatment and the manipulation of day length by artificial light regimes.

Intravaginal sponges or prostaglandin treatment may be used to induce ovulation. When used out of season, the withdrawal of the sponge is preceded by an injection of pregnant mare serum gonadotrophin (PMSG) two days earlier. The dose of PMSG is increased the further away from the natural season and the higher the milk yield of the animal. The procedure is summarised in Table 5.1. It is best suited to those times of the year when artificial manipulation of day length is difficult to achieve, i.e. when natural day length is longest. Most farmers therefore use sponges between June and October to advance the natural breeding season.

Controlled lighting is a simpler technique than the use of intravaginal sponges. The technique utilises the fact that the onset of oestrus in seasonal breeders is triggered by a reduction in the day length.

Table 5.1

**Out-of-season breeding may be achieved
by either:**

Hormone treatment:

Step 1	Insertion of progesterone-releasing intravaginal sponges for 21 days
Step 2	Injection of 400 to 600 i.u. pregnant mare serum gonadotrophin on day 19
Step 3	Females ovulate 24 to 30 hours after sponge removal

or

Controlled lighting:

Step 1	Expose females to 20 hours light per 24 hours for 60 days
Step 2	Reduce day length to 12 hours
Step 3	Females commence cycling 7–10 weeks later

Goats start their natural breeding season some 8 to 10 weeks after the longest day of the year. Most goat farmers choose to commence controlled lighting in the short days of January by exposing females to 20 hours of light. This is achieved by installing fluorescent lights controlled by an electric timer with a manual override. The lights should be of sufficient number to simulate daylight – about 1 foot of 40-watt tubing for each 10.5 square feet of floor space, mounted about 9 feet above the floor. The lights are timed to be on for 20 hours, since continuous 24-hour lighting does not produce the desired effect. To reduce energy costs the lights are effectively on from one hour before sunset until 1 am, and are switched on again from 5 am until one hour after sunrise. This regime is continued for 60 days. The actual start of the period depends on the desired time to begin autumn kidding. When the lights are switched off at the end of the period the day length is reduced to the natural period for the time of year – about 12 hours. Females commence cycling some seven to ten weeks later, in mid-May. The expected conception rate, with exposure to the male from mid-

April to mid-June, is 80 per cent. The procedure is summarised in Table 5.1.

As with natural breeding, the weight and body condition of the female are important criteria for successful conception. Flushing (see page 31) prior to commencement of the breeding programme is advisable to ensure a high proportion of females are ovulating normally.

Male fertility

Males are extremely active during the breeding season, and lose weight as a result. It is important therefore that they are flushed in the same way as females (see page 31) before the season commences.

Plate 4 A male in good breeding condition. (*courtesy of A. Mowlem.*)

It is a good idea to have males examined thoroughly before exposing them to cycling females. They should be anatomically sound, free of worms and in good body condition – though not overfat.

To avoid fighting it is advisable to keep males apart from each other when they are adjacent to or mixed with females.

During the breeding season a male would be expected to serve females at an average rate of one per day. Thus the normal ratio of males to females is 1 to 20. But the ratio should be narrowed if a high proportion of females have been synchronised, and may be widened if the breeding season extends for more than three months.

Artificial insemination

The most obvious advantages associated with AI are those which have been demonstrated in the dairy cattle industry where, for many years, large herds have been bred almost exclusively by artificial insemination. The main benefits (Table 5.2) are those which accrue from the wider choice of breed and type of sire and the exploitation of sires of superior genetic merit. In addition, the costs associated with keeping males are reduced and semen is available for trading on an international basis. This is a particular advantage where the movement of livestock between countries is not possible due to disease restrictions.

The disadvantages (Table 5.2) revolve around the management of the semen and the insemination. Sperm death can occur if freezing occurs in sub-optimal conditions. Correct insemination technique is important, particularly, in relation to timing and deposition of semen in the uterus.

Table 5.2 Pros and cons of AI for goats

Pros

- Wider use of top quality, progeny-tested males
- Wider choice of sires
- More convenient to transport semen than animals both within and between countries
- Well suited to synchronised and out-of-season breeding
- No need to keep males
- Blood lines can be preserved after the death of the male

Cons

- Reduced conception rate compared to natural mating
- Special equipment and skills are required
- Cost of semen and of insemination can be high

Semen, collected from males once or twice daily during the breeding season via an artificial vagina, is examined visually for viable sperm.

The normal density of sperm is about 2000 million per ml, with an average ejaculate of 1.2 ml. Size and sperm count of ejaculates vary with age and breed of male, and with time of year.

Once collected and examined, the semen is usually washed to removed seminal fluid and the sperm are re-suspended in glycerol to prevent the formation of crystals when frozen. The diluted semen is then placed in plastic straws. Depending on the size and sperm density, between 12 and 60 straws may be obtained per ejaculate. Straws contain either 0.25 or 0.5 ml of semen diluted to 250 to 600 million sperm per ml.

Once the straws are filled, their ends are closed and they are gradually cooled to 4°C before being plunged into liquid nitrogen at −196°C, where they are stored until shortly before use.

Thawing precedes insemination, which should occur towards the end of the oestrus period when the vaginal discharge is clear but beginning to thicken. Some organisations recommend a double insemination − the first, 12 hours after detection of oestrus, and the second, 12 hours later.

The female to be inseminated is first placed on a stand so that the vulva is at eye level, or she is held on the ground with her rear end raised. The vulva is cleaned and a sterile lubricated speculum inserted into the vagina. The cervix is located using a torch, and a test tube with a hole cut in the bottom, or a pipette, is gently inserted into the vagina. The thawed straw is held in a special catheter which is inserted into the cervix via the test tube in the vagina. The semen is deposited if possible in the uterus. If the cervix cannot be penetrated the semen should be deposited in the middle of the cervix rather than in the vagina.

Care has to be taken to avoid damaging the tender tissues of the vagina and cervix. It is also important to maintain the sterility of the apparatus to eliminate the risk of cross-infection.

Typical kidding rates (Table 5.3) in countries where large numbers of goats are inseminated, such as the USA, France and Switzerland, compare very favourably to that of 0.50 to 0.60 achieved with cattle.

The formation in the UK of Caprine and Ovine Breeding Services (COBS), with the remit to establish a national service for goat farmers, is a very welcome development. During 1984 and 1985 over 200 goats were inseminated. Approximately 70 per cent conceived to first insemination during the natural breeding season and 45 per cent conceived to first insemination after sponging out-of-season.

Hopefully, COBS will rapidly develop a service which offers the farmer a choice of semen from progeny-tested males of several

Table 5.3   Typical kidding rates in countries where artificial insemination is widespread

	Proportion of adult females kidding after AI at:	
	Natural oestrus	Synchronised oestrus
Raw, undiluted semen	0.75	0.65
Frozen, diluted semen	0.65	0.60

Source: Tervit, H.R., and Goold, P.G., personal communication 1982

different breeds, including not only males selected for high milk yield, but also others selected for their ability to produce high yields of fibre and carcasses of superior meat yield and conformation.

Embryo transfer

The technique of transferring embryos from a superovulated donor to a number of recipients is now well established in cattle. However, the technique has not been used on a commercial scale with goats, with the exception of New Zealand and Australia, where increased interest in goat fibre production prompted a demand for the rapid multiplication of Angora goats, using feral goats as recipients.

In a trial in New Zealand carried out over three years, 35 Angora donors were superovulated using PMSG and oestrus was synchronised using intravaginal sponges. Recipient feral goats, similarly synchronised, gave higher pregnancy rates when they received two embryos rather than one, and when the embryos were fertilised by young sires.

The donor goats were mated naturally shortly after removal of superovulated embryos. They produced a total of 154 kids from transferred embryos and natural mating (Table 5.4) – an average of 4.4 kids per female. This rate of reproduction is about five times greater than normal and illustrates the potential of the technique for rapid breed multiplication.

A further development is to split the embryos after their collection so that almost four times as many are available for transfer to recipients. This procedure is particularly well suited to use with very rare animals as donors.

Table 5.4   Embryo transfer in goats

Number of donors superovulated	35
Per donor:	
eggs collected	6.8
fertile embryos	6.0
transferable embryos	5.5
kids born in recipients	3.4
kids born from natural mating	1.0
Total kids per donor per breeding season	4.4

Source: Goold, P.G., and Tervit, H.R., personal communication 1982

Pregnancy diagnosis

Three methods may be used to establish whether or not a female is pregnant. The first is to observe the animal for signs of oestrus at 21 days and at 42 days after mating. If there is no return to oestrus the animal is presumed to be pregnant. The second method is to determine the concentration of progesterone in milk at 19 days after mating. The technique relies on the fact that the concentration of progesterone typically decreases sharply in the four days preceding oestrus and ovulation, and does not decrease if the animal is pregnant. It is actually a more reliable test of non-pregnancy than of pregnancy. Thus a high progesterone level in milk at day 19 is not necessarily followed by the birth of a kid because of the possibility of subsequent abortion.

The main advantage of the test is that if the female is diagnosed as being *not* pregnant at day 19, preparations can be made to re-inseminate her at or around day 21 when oestrus may or may not be observed.

The third approach is to confirm pregnancy by ultrasonic scanning. This technique is gaining widespread acceptance in sheep, and may also be suitable for use with goats. Since the technique involves identifying the foetuses, it is only suitable for use in mid-pregnancy when they have reached a sufficiently large size to be seen. The main value of scanning is that it can determine the number of foetuses being carried by the female and their stage of development, so that her level of nutrition in late pregnancy can be adjusted accordingly.

Pseudo-pregnancy ('cloudburst')

Occasionally females which have not been mated during the natural mating season will show a loss in milk yield, which commonly occurs once pregnancy is established. The goat will appear to be pregnant – the body will increase in size and will soften around the tail. Suddenly, around the normal full term of pregnancy (152 days), between 10 and 15 litres of a yellow watery fluid flood out but no kids are born. This 'cloudburst' or pseudo-pregnancy has no known cause and the goat returns to her former slim condition once the outpouring of fluid has occurred. Fortunately the majority of animals can be remated at the first oestrus after the discharge has completely cleared up and in all probability will breed normally.

Kidding

The most obvious signs of an impending birth are the rapid expansion of the udder and a slackening of the flesh around the vagina. A colourless discharge also appears from the vulva. The animal may appear distressed and refuse food.

After the onset of labour a membrane bag will appear from the vagina with the first signs of the kid. The normal presentation is two feet followed by the nose. A breech presentation, i.e. tail first, will require intervention to retrieve each leg before the kid is delivered.

The placenta or afterbirth should be discharged within two hours of the last birth. Retention of the placenta for longer than 12 hours requires prompt veterinary assistance to ensure that the entire tissue is removed and infection is avoided.

The umbilical cord of the kid should be treated with iodine to prevent infection and assist its rapid drying out. The newborn kid should be encouraged to drink about 100 ml of colostrum – either via bottle or by suckling – within the first four hours of its life.

Rearing kids to weaning

Kids should remain with their mothers for two days to ensure they receive adequate colostrum. This acts as a vital source of antibodies to protect the young animal from infections, and also acts as a laxative to stimulate the alimentary tract to function. Some goat farmers prefer to bottle-feed heat-treated colostrum to ensure that viruses and bacteria are inactivated.

After the first two days kids may be artificially reared using milk replacer. Proprietary milk powders specifically designed for goats may be used, but good results have been obtained with calf milk replacer, which is cheaper (Fig. 5.1).

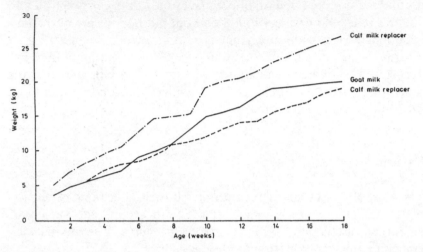

Fig. 5.1 Growth of kids from 1 week to 18 weeks, fed goat milk, or calf milk replacer reconstituted at two concentrations. Source: Mowlem (1984) Artificial rearing of kids *Goat Vet. Soc. Journal 5*: 25–30.

After removal from the mother, kids can be reared in pens of about 12 animals. Milk is normally given by bowl, or bottle, or a teat supplied from a central reservoir. If milk is restricted, the animals should receive about 750 ml in three daily feeds up to 6 weeks of age. Clean water should be available at all times. Good quality hay and a coarse mixed concentrate feed should be offered from two weeks of age. Weaning occurs from 8 to 10 weeks of age when the animals have established a regular intake of solid food (see Chapter 8).

Early kidding

Rearing for early kidding usually involves giving females born in January and February a carefully controlled diet so that they gain in weight continuously to reach about 60 per cent of their adult weight at nine months of age, i.e. 40 to 45 kg. The intention is to mate them in their first year of life so that they kid at about 15 months of age.

Plate 5 An automatic milk feeding system for rearing goat kids. (*Courtesy of British Denkavit Ltd.*)

Further reading

Anon (1985) 'Smallholder's guide to successful kid rearing' *Goatkeeper and Smallholder*, **3** (8) March 1985, 12–14.

Mowlem, A. (1982) 'An artificial insemination scheme for goats in the UK' *Goatkeeper and Smallholder*, **1**: 24.

Mowlem, A. (1983) 'The development of goat artificial insemination in the United Kingdom' *British Goat Society Yearbook 1983*, 4–6.

Tervit, H.R., Goold, P.G., McKenzie, R.D. and Clarkson, D.J. (1983) 'Techniques and success of embryo transfers in Angora goats' *New Zealand Veterinary Journal,* **31**: 67–70.

6 Health

The majority of commercial herds are kept indoors all the year. Whilst this reduces problems of infection with nematode worms at pasture, other hazards such as enterotoxaemia, pneumonia and coccidiosis can present greater problems. It is especially important that the health of the whole herd is managed in an organised way so that diseases are prevented rather than cured, particularly those which can be passed rapidly between closely confined groups of animals. This means devising and implementing a herd health scheme.

Herd health scheme

Regular inspection of the herd, with timely intervention, vaccination and culling, are the essential elements of a successful herd health scheme. Normally, the local veterinary surgeon is consulted and a programme of regular visits is drawn up. The objective is to concentrate on preventative medicine so that losses due to illness or mortality are minimised. At each visit the performance of the herd is examined, individual animals are inspected, routine vaccination undertaken and treatment, if necessary, is carried out.

The frequency of visits by the vet may vary, but with large herds of more than 250 animals it is advisable to have a visit once every two weeks and preferably once each week throughout the year.

In addition to attending to the health of the herd, the vet can assist in confirming pregnancy, and advise on a suitable culling policy for those animals whose performance is below target.

Infectious diseases

Goats share susceptibility to many of the diseases of sheep. Thus it is wise policy to adopt a programme of vaccination early in the goat's life, as with lambs. It goes without saying that all new arrivals on the farm should be isolated for the first few weeks.

It is not the intention of this book to describe comprehensively all the infectious diseases of the goat, but rather to draw attention to those which are most likely to be encountered in the management of a commercial herd. Thus the diseases discussed here are those associated with keeping stock in confined areas, those of intensive grazing, and those of the mammary gland.

Enterotoxaemia

This is a disease to which goats appear to be very susceptible at all ages. The disease appears suddenly, the animal becomes lethargic, and unable to stand, and may develop a watery diarrhoea. Death often follows quickly. Vaccination against enterotoxaemia should be carried out twice a year. Immunity can be conferred on the kid by vaccinating the mother two weeks prior to kidding. The kids themselves are vaccinated at 10 to 12 weeks of age and again six weeks later, with further vaccination every six months.

The disease can often break out following a change of diet.

Caprine arthritis-encephalitis

This is caused by a retrovirus which infects the brain and central nervous system. The encephalitis form usually affects kids between two and four months old. Symptoms comprise head tremors, loss of co-ordination, and partial paralysis leading to recumbency and death. The arthritis form of the disease occurs in mature goats and affects the feet and other limb joints. Other symptoms are loss of condition and poor hair coat. 'Hard udder' is also a common symptom. This is a condition in which the udder is very hard after kidding, and little or no milk is produced. The major route of transmission of the virus is to young kids through infected colostrum and milk, but direct contact between infected and non-infected goats can transmit the disease.

There is no effective treatment for the disease. At present the prevalence of infected animals is quite low (Table 6.1) but failure to take precautions to limit the spread of the disease both within and between herds could lead to a rapid increase in the number of infected animals.

Table 6.1   Prevalence of caprine arthritis-encephalitis in British goats

Number of herds surveyed	331
Average herd size	8
Percentage of herds with infected animals	10.3
Percentage of infected animals	4.3

Source: Dawson, M. and Wilesmith, J.W. (1985) *Veterinary Record* **117**, 86–89

The best method of preventing infection is to test suspected adults. If they are carrying the virus they should be culled from the herd forthwith.

Mastitis

Mastitis is a bacterial inflammation of the mammary gland, which can be difficult to diagnose in goats, especially if it is of the non-haemolytic type and the infection is at a low level.

The most common symptom of mastitis is clots in the milk, but not all clots in the milk are caused by mastitis – occasionally small pieces of mammary gland tissue are sloughed off during milking; this condition is quite normal, and reflects the higher content of tissue cells which is often recorded in goat milk compared to that from cows. Tissue cells look like small clots of mastitic milk but they feel rubbery and do not disintegrate on pressing between the fingers as mastitic clots do.

Mastitis can be caused by a variety of bacteria entering the teat canal. It is usually a reflection of poor udder hygiene, poor milking technique, poor udder conformation or injury to udder and teats. Treatment is as for mastitis in cows. The udder is stripped out, the milk discarded, and an antibiotic is administered either locally via the teat orifice or systemically.

Chronically mastitic goats should be culled regardless of their milking ability since they act as a reservoir of infection in the herd.

Surveys of udder infections have shown that the milk from uninfected udders contains very low levels of bacteria, less than 100 per ml of milk (Table 6.2). However, the milk of goats has a characteristically high content of total cells. In one study, cell count was similar

Table 6.2 Percentage distribution of total bacterial count (TBC) in infected and non-infected samples of milk

	TBC/ml of milk		
	<100	100 – 1000	>1000
Infected samples	9.1	10.7	80.2
Non-infected samples	79.5	16.8	3.6

Source: Hunter, A.C. (1984) *Veterinary Record* 114, 318–320

between uninfected halves of the udder and halves infected by coagulase negative staphylococci (Table 6.3). Thus coagulase negative staphylococci appear to be constituents of the natural flora and not associated with clinical problems. Coagulase positive staphylococci on the other hand, are pathogenic, and it appears that *Staphylococcus aureus* is the major pathogen responsible for mastitis in goats.

Table 6.3 Udder infections and somatic cell counts in French goat herds

Number of herds surveyed	10
Average herd size	61

	Milk samples		
	Contaminated by major pathogens[1]	Contaminated by coagulase negative staphylococci	Not infected
Percentage of milk samples	7.5	23.8	68.7
Somatic cell count cells × 10^6/ml	6.77	1.78	1.54

(1)   *Staphylococcus aureus* (75%), streptococci (23%)
Source: Lerondelle, C. and Poutrel, B. (1984) *Annales de Recherches Vétérinaires* 15, 105–112

A cell count in mid-lactation exceeding a threshold level of 1×10^6 cells/ml of milk can reliably indicate the presence of major pathogenic infection in the absence of clinical symptoms of mastitis.

Dry goat therapy is the best way of reducing the level of mastitis in the herd. The advantages over lactation therapy are that there is no milk to be discarded, longer-acting antibiotics can be used to achieve greater control of bacteria, and sub-clinical sources of infection are treated without the need to detect them in the first place. The procedure is to give antibiotic treatment to all goats after they are dried off, as routine practice.

A useful additional routine is to dip all teats in a sterilising solution at each milking. The objective is to prevent the passage of infection from teat to teat through the milking equipment.

Johne's disease

This is a wasting disease caused by *Mycobacterium paratuberculosis*, a small rod-shaped bacterium. The most obvious symptoms are inappetence and progressive emaciation over several months, caused by a thickening of the wall of the intestine and reduced absorption of water and nutrients. Diarrhoea may develop a few days before death.

The bacteria may be transmitted through the uterus, in milk, or by faecal contamination of feed, water or bedding. There is no effective treatment and it is vital, once infection is present in a herd, that the herd is checked regularly to identify animals which may be carrying the bacteria. This can be done by culturing samples of blood or faeces. Any wasting animals must be checked for the disease.

If an outbreak of Johne's disease occurs, kids should be isolated from the rest of the herd and all stock should be rigorously examined every six months until all positive animals have been identified and culled.

Listeriosis

A disease which is usually linked to the use of silage which has been contaminated with soil or exposed to air during storage or during the feed-out period. It is caused by the bacterium *Listeria monocytogenes* which can grow in silages of relatively high pH (greater than 4.5) in the presence of air.

The risk of listeriosis is likely to be greater with big bale silage than with silage made in bunkers or clamps. If big bales are to be made into silage it is essential that they are completely sealed at the outset and stored so that the bags are protected from damage by wind, birds, vermin, animals or children. With all silages soil contamination must be as low as possible.

Clinical signs are depression, elevated temperature (above 40°C), partial facial paralysis leading to circling behaviour with a head tilt, loss of appetite and death. If the animal is pregnant, abortion occurs after the twelfth week. Treatment is by individual antibiotic therapy but the prognosis is usually poor.

Pneumonia

Pneumonia is normally a reflection of poor ventilation, either in houses or in transit. Goats appear to be very susceptible to damp, draughty, ill-ventilated conditions. One of the most common causative organisms is *Pasteurella haemolytica* which may cause death following symptoms such as coughing with nasal discharges, listlessness and fever. Occasionally the only symptom is sudden death. Antibiotic therapy, if given at the earliest stages of the disease, can be successful and is a wise precaution if one animal in a group has succumbed to the disease.

Prevention of pneumonia is achieved by improving the ventilation so that air changes above the animals are more frequent, and draughts are eliminated. Bedding should be kept as dry as possible. Vaccination against *Pasteurella* is a worthwhile precautionary measure, and an essential part of a herd health scheme.

Abortions

Abortions are relatively uncommon in goats. They can be caused by infections, including *Toxoplasma*, which is a public health hazard since the organism can also infect pets and humans. Treatment of toxoplasmosis is by antibiotic administration.

Parasites

Coccidiosis

This can be common in herds which are housed all year round. Infection may be present at a low level in older goats and this can serve as a source of more serious disease in kids. Symptoms include poor condition and diarrhoea. Control is by antibiotic treatment, and prevention by ensuring that young kids are not kept on the same bedding as that used for older animals.

Gastrointestinal helminths

These can infect goats in a similar manner to sheep, particularly if the herd is allowed to graze infected pastures. Small paddocks close to the buildings can act as a reservoir of infection from year to year, necessitating repeated anthelmintic treatment of the herd.

A system of 'clean' grazing, advocated for sheep flocks, can break the annual cycle of the worms and their larvae, and reduce the need for continual worm treatments. The principle is to alternate grazing by goats with either cutting or grazing by cattle. Since goats can be susceptible to some cattle parasites the best plan is to cut the fields in alternate years, or to worm the animals immediately before turning them onto clean regrowth areas. In addition, the herd should be wormed routinely before being turned out to grass so that clean animals graze clean pastures from the outset. It is important to avoid the build-up of infection which often occurs in spring (the so-called 'spring rise'), as worm eggs shed from the grazing animal are added to the eggs which remain on the pasture from the previous grazing season.

Goats show a rise in faecal egg counts around the time of kidding. This contaminates the bedding and acts as a source of infection for the new generation of kids. It also reduces milk production. Goats should therefore be wormed at the beginning of the winter period so that the level of infection during housing is kept low. Kids should also be wormed at six weeks of age to avoid a serious build-up of infection which may result from the ingestion of worm eggs or larvae early in life.

The most obvious symptoms of worm infestation are poor condition and diarrhoea. Treatment is achieved by anthelmintic administration, usually in the form of an oral drench, though injectable preparations are also available.

The most common worm species are *Haemonchus*, *Ostertagia*, *Trichostrongylus*, *Nematodirus* and *Cooperia*.

Nematodirus infection is particularly dangerous in kids, as in lambs. It can cause rapid death early in the grazing season. This occurs because the infective larvae develop inside the eggs during the winter period and hatch early the following spring. Kids eat the larvae in large numbers and soon show clinical symptoms of acute diarrhoea, dehydration and death within a two-to four-day period.

It is important to ensure that kids are protected from *Nematodirus* infection by worming them at turnout and at monthly intervals in spring, and by avoiding the use of the same pastures or exercise areas which were grazed by kids the previous year.

Physiological and metabolic disorders

Physiological and metabolic data for the goat are given in Table 6.4.

Packed cell volume

This is closely related to blood haemoglobin content and indicates whether or not the animal is anaemic. Values tend to be lowest in early lactation and in late winter, and highest in late lactation and in mid-summer. When low packed cell volume is associated with low plasma urea levels, the diet is likely to be seriously deficient in protein.

Total protein

The total protein in serum is the sum of albumin and globulins. Albumin is determined together with total protein and globulin is estimated by difference. A high globulin level can indicate an immunological response to infection. A low concentration of albumin may indicate a deficiency of protein in the diet, or a damaged liver since albumin is synthesised in the liver.

Urea

Urea is a measure of the amount of ammonia produced in and absorbed from the rumen. High values can indicate either a shortage

Table 6.4   Physiological and metabolic data for goats

Parameter	Value
Temperature	39.4°C (38.6 to 40.6°C)
Heart rate	70 to 90 beats/min.
Respiratory rate	15 to 30/min.
Urine:	
pH	7.0 to 8.0
volume	10 to 40 ml/kg body weight/day
Semen:	
volume	0.5 to 2.5 ml
sperm concentration	1 to 5×10^9/ml
Gestation period	150 days
Blood values:	
packed cell volume	280 to 350 ml/l blood
total protein	60 to 78.5 g/l serum
albumin	32.5 to 49 g/l serum
globulin	23 to 46 g/l serum
ketones	100 mg/l blood
non-esterified fatty acids (NEFA)	226 µmol/1 plasma
glucose	2.2 to 3.3 mmol/l plasma
urea	4.0 to 9.9 mmol/l plasma
calcium	2.4 to 2.6 mmol/l serum
magnesium	1.0 to 1.4 mmol/l serum
potassium	3.5 to 6.3 mmol/l serum
chloride	100 to 120 mmol/l serum
inorganic phosphate	1.2 to 2.5 mmol/l serum

Sources: Lloyd, S. (1982) *British Veterinary Journal* **138**, 70–85; Ministry of Agriculture, Fisheries and Food/Department of Agriculture and Fisheries for Scotland (1984) *Blood Characteristics and the Nutrition of Ruminants Reference Book 260*, London: HMSO

of readily available energy (e.g. starch or sugar) in the rumen, or a high intake of non-protein nitrogen or of rumen degradable protein. High levels of plasma urea are commonly seen in animals grazing leafy spring or autumn grass. Lower values are more likely in the winter period, particularly if concentrates comprise a high proportion of the diet.

Glucose

Glucose levels indicate the energy status of the animal. Low values follow under-feeding and are often accompanied by low solids-not-fat values in milk. Animals in early lactation can also exhibit low blood glucose levels, particularly if they are giving high milk yields and are losing body weight.

Fatty acids and ketones

Non-esterified fatty acids and ketones reflect the mobilisation of body fat reserves to meet the energy needs of the animal. Thus higher values are more likely to be seen in the first few weeks of lactation.

Inorganic phosphate

Inorganic phosphate levels tend to reflect dietary intake of phosphorus, but levels may also increase where weight loss is occurring. Diets low in phosphorus may be reflected in low blood phosphorus concentrations and also in elevated levels of serum calcium. Low phosphorus levels can occur with major dietary changes – especially the change from winter feed to spring pasture.

Magnesium

Serum magnesium reflects magnesium intake. Supplementary magnesium in the diet, whilst increasing serum magnesium, can lower the level of serum phosphorus. Serum magnesium is also lowered by an excess of phosphorus or potassium in the diet; increased potassium intake by the animal results in a decrease in the absorption of magnesium.

Calcium

Calcium levels in serum are depressed when diets are high in phosphorus or low in calcium, but the effect may be temporary, since calcium levels are maintained by mobilisation of bone calcium.

Chloride

Chloride concentrations indicate the balance of electrolytes in the blood: elevated levels reflect dehydration.

Pregnancy toxaemia

Pregnancy toxaemia is seen in late pregnancy in goats which are in poor condition. Often, the disorder is precipitated by a sudden decrease in energy intake. The animal stops eating and becomes constipated. It may abort or give birth prematurely. As the condition progresses, the animal becomes uncoordinated, then comatosed. Treatment is by administration of glucose or dextrose followed by propylene glycol twice daily for two to three days.

Since pregnancy toxaemia is the result of the animal drawing on body reserves of fat to maintain nutrient supply not only for her own metabolism but also for foetal growth, abortion or parturition can alleviate the condition. Prevention of pregnancy toxaemia is relatively easy to achieve. As with the ewe, the two foetuses occupy an increasing amount of space as pregnancy advances. Thus the space available for the digestion of feed in the rumen is reduced. As a result intake of bulky feeds decreases during pregnancy. The reduced intake can predispose the animal to pregnancy toxaemia unless the energy concentration of the diet is increased to compensate. This is usually accomplished by increasing the level of concentrates during the last eight weeks of pregnancy (see Chapter 8).

Ketosis

Ketosis, like pregnancy toxaemia, is a disorder associated with an imbalance between nutrient supply and nutrient requirements. Metabolism in early lactation of fatty tissues laid down during pregnancy can give rise to elavated levels of ketones in blood, which can rise from 100 to 1000 mg per litre of blood. Thus the objective is to prevent the pregnant goat from becoming over-fat during pregnancy, and to prevent those animals which are giving high yields of milk in early lactation from being under-fed. The condition is often seen when goats are over-fed during pregnancy, with a diet high in carbohydrate and low in protein. Lack of exercise during pregnancy also contributes to the onset of ketosis in early lactation.

Symptoms are similar to those of pregnancy toxaemia, with the characteristic 'pear-drop' flavour of ketones in milk and on the breath. Treatment is similar to that for pregnancy toxaemia (see page 52), by administration of glucose or dextrose followed by propylene glycol. Many people feed diluted treacle to all animals for the first week or so after parturition.

Hypocalcaemia

Hypocalcaemia, otherwise known as lactation tetany, occurs around the time of kidding. The symptoms are muscle tremor leading to collapse of the animal. Heavy breathing and bloat can also occur. The disorder is usually a reflection of the sudden increase in demand for calcium following the onset of milk production. The animal is unable to meet the sudden surge in demand, either from the diet or by drawing on body reserves, and the level of circulating calcium in the blood drops as a result.

Treatment is by intravenous or subcutaneous injection of calcium borogluconate. Prevention is by avoiding the feeding of high levels of calcium during pregnancy, so that the animal is encouraged to mobilise her body reserves. In addition, adequate supplementary calcium should be included in the diet in early lactation (see Chapter 8).

Hypomagnesaemia

Hypomagnesaemia, or grass tetany, occurs following turnout to grass in spring and can also occur in autumn in certain areas of the country. Essentially, the disorder is a reflection of low blood magnesium levels. The only observed symptom may be sudden death. In the early stages the legs stiffen and the animal may look like a rocking horse. The goat may also stagger about, grind its teeth and collapse in a fit before becoming comatosed and dying.

Rapid treatment by subcutaneous injection of a magnesium solution can lead to rapid recovery. Prevention is achieved by including supplementary magnesium in the diet in spring and autumn. This may take the form of calcined magnesite or magnesium phosphate. Alternatively, magnesium acetate can be added to the drinking water or directly to the feed.

It is important to avoid the application of compound fertiliser containing potassium to grassland in late winter and early spring, since the potassium reduces the availability of soil magnesium to the plant and can lead to low concentrations of magnesium in herbage. In addition, if potassium levels are higher than normal in the herbage, this can impair absorption of magnesium by the animal.

Deficiencies and toxicity

Mineral and vitamin deficiencies and toxicity may result in a variety of disorders. Veterinary and nutritional advice is needed to diagnose and rectify the problem.

Poisoning

It is important to be aware that certain plants (e.g. ragwort) are poisonous to goats and it is worth checking that pastures to be grazed or conserved do not contain plants liable to cause poisoning.

Disbudding and castration

The horn buds of the goat are considerably larger in relation to the size of the head than in calves, and the nerve supply is more complex. Disbudding and dehorning must therefore be done by a veterinary surgeon who has experience of the practice. It is illegal in the UK for people other than veterinary surgeons to disbud goats.

Castration should be performed in the first week of life by use of a rubber ring to constrict the flow of blood to the scrotum. It is important to ensure that both testes have descended into the scrotal sac, and that the ring is placed above both testes.

Lameness

Lameness can be a major problem of housed goats, associated with the nature of the floor. Dry abrasive surfaces are ideal. Wet bedding and lack of exercise predispose to problems.

A high incidence of lameness may reduce herd milk yields.

Preventative measures include checking housing management, and using a footbath regularly. Vaccinations are also available against foot rot.

Movement records

MAFF stipulate that goatkeepers maintain a record of their animals' movements. This is to help trace contacts if there should be an outbreak of a notifiable disease (e.g. foot and mouth disease, anthrax, etc.).

Further reading

Anon (1986) 'How to have a healthy herd' *Farmers Weekly*, 16 May p38.

Dunn, P. (1982) *The Goatkeeper's Veterinary Book*, Ipswich: Farming Press Ltd.

Edwards, L.M. (1983) 'Behaviour and diseases of the dairy goat' *The Animal Health Technician*, 4: 294–300.

Cooper, M.R. and Johnson, A.W. (1984) *Poisonous Plants in Britain and their Effects on Animals and Man*, MAFF Reference Book 161 London: HMSO.

Grunsell, C.S.G., Hill, F.W.G. and Raw, M.E. (Eds) (1985) *The Veterinary Annual*, 25 Bristol: Scientechnica.

MAFF/ADAS (1982) *Some Veterinary Notes for Goat Keepers*.

MAFF/ADAS (1982) *Towards a Healthy Herd*.

Symposium on Sheep and Goat Medicine (1983) Veterinary Clinics of North America *Large Animal Practice* 5: Part 3.

Watson (1984) 'The import and export of sheep and goats' *British Veterinary Journal*, 140: 1–21.

7 Housing and Equipment

The provision of a suitable environment is clearly of fundamental importance to the success of a goat enterprise. The animals must have adequate space, light, ventilation, access to feed and water, and freedom to exercise. In a large herd, the welfare of animals which are sick, or about to give birth, must be considered. Therefore, in order of priority, the following points should be taken into account in housing and equipping a commercial herd:

1. healthy animals
2. a high standard of animal welfare
3. hygienic milk production
4. suitable siting of buildings
5. suitable layout of facilities
6. appropriate equipment
7. economy in cost of buildings and equipment
8. plans for the future

The housing, environmental and feeding facilities should suit the animals rather than the people who have to manage them. The milking facilities, on the other hand, can be designed to suit the stockperson.

An efficient milking system assists in the welfare of the animal and eases the chore for the milker. Milking should rarely take longer than 1½ hours, otherwise the standard of work tends to decline.

Goats are great climbers: they have been known to climb up trees to reach leaves! Housing and equipment must therefore take account of the fact that anything lower than 2 metres above the ground is liable to be chewed or, if it is a sharp projection, liable to hurt the animals. All light fittings and switches should be protected by wire mesh. Exposed plastic pipes should be avoided. Walls should not be painted. Timber should be clad in metal sheeting. Partitioning should be at least 1.2 metres in height and preferably 1.5 metres high, especially if the herd contains Anglo-Nubians.

Basic facilities

Floors

The choice is essentially between a slatted floor or one which is bedded with straw or wood shavings.

SLATTED FLOORS

Slats have the advantage that the animals' feet are likely to stay in good condition, and the animals themelves usually remain clean. A greater stocking density is possible on slats than on bedded floors. The ability to accumulate manure in a pit below the slats reduces the frequency of the need to remove solid manure and the problem of a rising level of litter. Slats normally remain dry; the labour input is low, and there is no requirement for straw.

The disadvantages of slatted floors are that the capital cost of installation is high, and there is risk of injury if a slat breaks. Waste feed may block the gaps and in many cases they require regular maintenance to work successfully.

Slats may comprise wood, wire, steel mesh, expanded metal or concrete. Whatever material is chosen, it should be slip-resistant, self-cleaning and constructed to allow easy removal of manure. The recommended width of slatted floors is 25 to 100 mm per slat with a gap between slats of 16 mm. The minimum depth of the manure pit below the slat is 1 metre.

Space per animal is fairly critical with slatted floors; around 0.8 to 0.9 square metres per adult animal should ensure the animals and the slats remain clean.

STRAW, SAWDUST AND PEAT FLOORS

By far the most popular floor is one of concrete or hard core covered with straw bedding. Whilst a greater floor area is required per animal for straw than for slats, capital cost is much lower. However, mucking out requires more time than with slatted floors. It is essential that bedded floors are kept relatively dry or the incidence of foot disorders can increase.

Straw, sawdust and peat all act as insulation and thus provide a warmer winter environment for the animals. Goats usually pick over fresh straw after it is put out, and some farmers deliberately use straw for feed and bedding, as a source of long fibre in the diet. The normal requirement for straw bedding is between 200 and 400 kg per animal for a six-month winter period.

Sawdust and wood shavings are occasionally used as bedding but sawdust in particular can ball up in between the cloves of the feet and cause increased lameness. Peat tends to be expensive as a source of bedding.

As the level of bedding rises there is increased risk of animals escaping or being unable to reach low-lying feed troughs outside the pens. Regular removal of manure every two to three months is therefore essential.

Exercise areas

If the herd is to be kept indoors all year round, the animals require access to an exercise area. If it is a grass paddock, the area will be very heavily stocked indeed and it is wise to budget for no grazing. It can act as a very fertile breeding ground for parasitic infections (see Chapter 6) and regular worming is necessary if the same paddock is used daily throughout the year. It is worth having separate areas for adults and young stock to avoid cross-infections. The minimum exercise area necessary is 1.7 square metres per adult goat.

Water supply

A reliable supply of clean, fresh water is vital, preferably delivered to the animal through easily-cleaned, self-filling troughs. Goats tend to dislike water bowls. Troughs should be easily accessible and away from sources of contamination such as feed and manure.

Feed supply

The most important feature of the feed supply is that it must be accessible to all animals but not too accessible so that animals can climb into racks and troughs to contaminate the feed or escape. Overhead racks can supply hay and straw, and troughs or a floor passageway with feeding bars may be used for silage, concentrates and by-products.

Feed storage should be under cover, dry and protected from vermin, and also from animals which may escape from their pens. Adequate feed storage allows economies in the cost of feed because bulk discounts can be secured.

Types of housing

There are four main types of housing – individual stalls, individual pens, cubicles and bedded yards. The space requirements for adult females for the four systems are given in Table 7.1.

Table 7.1 Space requirement for adult female goats

	Minimum space required (m^2 per goat)
Individual stalls	0.5
Slatted floor pens	0.9
Individual pens	2.0
Cubicle	2.4
Bedded yard	1.7

Individual stalls

Individual stalls have a low space requirement per animal, high labour requirements and relatively high construction costs. An exercise area is also needed.

Individual pens

Individual pens, like stalls, have the advantage that each animal can be isolated. Pens allow more room for exercise but in consequence the space requirement is much higher than for stalls. Construction costs are higher than for stalls or yards and they have a high labour requirement.

Cubicles

Cubicles should be constructed to match the size of the goat, to avoid problems of lameness and of animals lying in passageways. They are well suited to farms where the design of buildings and layout of fields does not allow easy access to an exercise area.

Bedded yards

Bedded yards are by far the most popular housing for larger goat herds. The space required is moderate and an exercise area, whilst useful, is not essential. Construction costs can be low and the system is adaptable to a wide range of herd sizes and feeding systems. However, the animals are not kept as individuals and additional isolation pens are necessary.

The optimum number of goats per yard depends on the size of the herd and the milking parlour. Animals should be grouped according to stage of lactation. Ideally, the group size of each yard, or of each pen in a yard, should be equal to or a multiple of the number the parlour holds at one time.

Isolation pens

Isolation pens should preferably be in a separate building from that which houses the herd. Each pen should be self-contained, able to be thoroughly cleaned, dry, warm, and have adequate ventilation. It should be possible to milk in each pen. Access to each pen must be good so that sick or injured animals can be moved in and out with relative ease.

Kidding pens

Kidding pens are a useful accessory but are not vital if the herd is housed in bedded yards. Goats can kid in the yard but it is always useful to be able to separate the newly-kidded goat and her newborn offspring from the rest of the herd. It is easier to tend to the animals and check them if they are in a special pen, which may simply be temporary hurdles erected in the corner of a yard. About 2 square metres per goat should be allowed for each pen.

Designs of bedded yards

There are three basic designs of yard, as shown in Fig. 7.1.

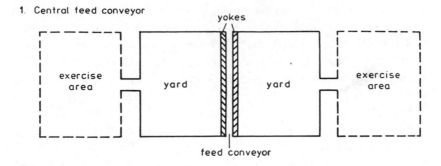

1. Central feed conveyor

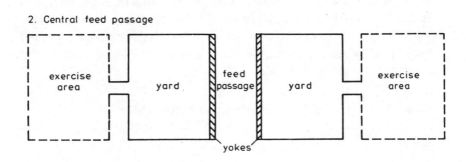

2. Central feed passage

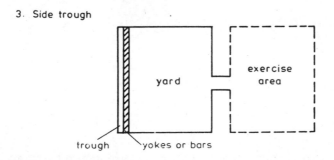

3. Side trough

Fig. 7.1 Basic designs of yards for goats.

Central feed conveyor

This design is well suited to feeding systems in which feeds are mixed together, or for conveying forage mechanically to the animals. Silage or zero-grazed herbage may be directly unloaded onto the conveyor which moves the feed slowly through the yards. Rejected material is discharged at the opposite end. The design has been popular in France where more recent installations comprise two conveyors, each 0.5 metres wide with a middle passage (1.2 metres wide) down which the stockperson can walk to inspect the animals. A mini-trough above the conveyor may be installed to hold the concentrate part of the diet.

Central feed passage

The central feed passage may comprise a floor-feeding system or troughs, with yokes or feeding bars along the sides. The passage is often constructed so that a forage box, or other bulk feed unloader, can be driven down the passage to deposit feed within reach of the animals. Big bales or blocks of silage can be dropped in the passage and feed distributed by hand as required. The minimum widths of passage and headroom are given in Table 7.2.

Table 7.2   Central feed passage: minimum widths and headroom

	Width (m)	Headroom (m)
Wheelbarrow	1.5	2.0
Tractor (no cab) + trailer	3.0	2.2
Tractor (+ cab) + trailer	3.0	3.0
Tractor (+ cab) + forage box	4.5	3.0

Side trough

This is often constructed as a low-cost way of housing and feeding a larger group of animals where space is limited. The trough may be outside the building with access for a tractor-driven forage box or feed dispenser. Purpose-built side-trough yards usually have an overhanging roof to protect the trough from the weather. Access to the trough is via yokes or feed bars.

Feed barriers, troughs and racks

The two most common designs are diagonal feed bars and key hole yokes. They are shown diagrammatically in Fig. 7.2

The principle of both designs is that the animal is discouraged from withdrawing its head through the barrier before it has finished eating a mouthful of feed. If bars are used, the slope must be sufficient to prevent animals escaping.

Troughs may be almost non-existent if feed is placed on the floor of a central feed passage, but it is necessary to have some means of preventing animals from pushing their feed out of reach; a small upright concrete slab may serve the purpose.

Plate 6 Raised feeding troughs positioned outside the loose-housing pen. This arrangement prevents goats from climbing into the trough and contaminating the feed. (*Courtesy of British Denkavit Ltd.*)

(a) Diagonal bars

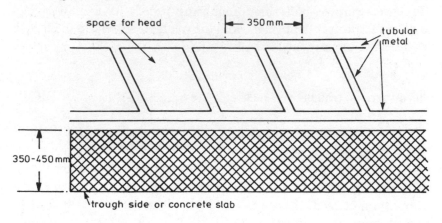

(b) Keyhole yokes

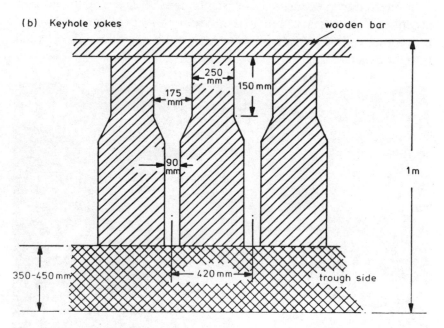

Fig. 7.2 Common feed barriers.

Alternatively, a complete trough may be constructed. Minimum trough dimensions are:

length = 0.4 m/goat
width = 0.4m
depth = 450 mm

Racks for hay should be about 1.25 metres above the bedded area. Close bars or coarse wire mesh will help to reduce wastage. The racks should be closed at both ends to prevent animals from jumping into them and fouling the feed, escaping, or breaking a leg by trapping it in the rack.

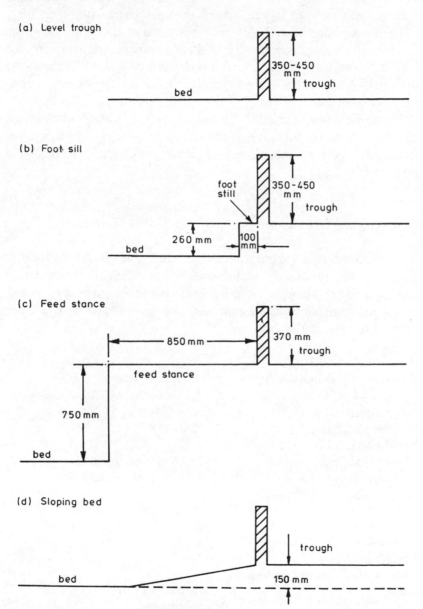

Fig. 7.3 Basic designs of trough, feed barrier and bed.

There are four basic designs of bedded area, as shown in Fig. 7.3. If the bedded area is on the same level as the feed trough (Fig. 7.3 (a)) the construction cost is low, but the design has the disadvantage that the trough becomes progressively lower as the bedding builds up.

The foot still (Fig. 7.3(b)) ensures that the animal retains the same position with its front legs relative to the feed barrier and trough.

The feed stance (Fig. 7.3(c)) may be of variable height above the floor of the bed. Theoretically it should remain relatively free of bedding so that the animal's position remains constant relative to the trough. In practice it is likely to require cleaning occasionally to keep the level constant.

The sloping bed (Fig 7.3(d)) may also require occasional cleaning. The slope should be sufficient to prevent bedding from building up, but not so steep that the animals have difficulty walking up it to reach the feed trough.

Environmental factors

The optimal housing environment for adult goats is summarised in Table 7.3. Essentially, the building should have adequate ventilation from spaced boarding or plastic netting between the eaves and the wall so that the warm air from the animals moves upwards towards an open ridge at the top of the roof.

Table 7.3   Optimal environment for housed adult goats

Temperature	10 to 20°C
Relative humidity	70 to 75%
Ventilation:	
air changes	7 to 8 m³/hour/animal
maximum air speed	0.5 m/sec. at animal level
Air flow above animals:	
winter	30 m³/hour/animal
summer	120 to 150 m³/hour/animal
Wall height	2.1 m
Eaves height	3 m
Open ventilated ridge cappings	
Light	natural light from 5% of total covered area

Cooler, fresh air should circulate around the animals after entering above the walls. The height of the walls is important, to avoid draughts on the animals. The building should be sufficiently well ventilated so that the doors may be kept closed in all but the hottest weather.

Large daily fluctuations in temperature should be avoided if possible. The bedding will serve as a source of heat in addition to the animals, so that in a draught-free building, stocked correctly, the temperature should remain above the minimum of 6°.

Windows, spaceboards or roof lights should allow natural light to enter the building from 5% of the total covered area. Artificial light is essential for the inspection of stock at night, and may be required for out-of-season breeding (see Chapter 5).

Males

The essential features of housing for male goats are:

- well away from the milking parlour
- strongly-constructed individual pens or grazing paddocks
- visual contact with other goats

Pens or paddocks should be of sufficient size to allow the animals adequate daily exercise to avoid lameness. Visual contact with other goats is important – if the males are housed near the females then oestrus is more likely to be stimulated and observed. But to avoid the risk of tainting milk it is essential to house males well away from the milking parlour.

Pens should be larger than for females – at least 2.3 square metres. The exercise paddock or yard should have walls or strong post and rail fencing with plain wire between the rails. The height of walls or rails should be at least 1.5 metres for Saanen or Alpine males, and 1.75 metres for Anglo-Nubians.

If the male is tethered, the tether must be moved and inspected regularly. It should be sited away from other goats, be strong and correctly designed with a swivel at each end and in the middle.

Young stock

Environment

The environment should be similar to that for adults (see page 66), but a source of heat should be available for pens of very young kids. Fans may be required to assist the natural ventilation, especially if an existing building has been converted to house the animals.

If possible, kids should be housed in groups of similar age and size, with about 12 per group the optimal size. It is preferable to divide the house into pens using solid partitions, to reduce the risk of cross-infection by physical contact between individuals in adjacent groups.

The size of pens can vary, and it is useful to be able to subdivide the young stock building with portable barriers so that the size of each pen may be varied to suit the number in the group and their age. Each pen should be capable of being emptied, thoroughly cleaned and rested between batches.

Design of pens

Several designs of pens are possible; they are illustrated in Fig. 7.4.

Single-row pens are most common in narrow buildings which have been converted for rearing young stock. The two-row pen layout with central feed passage is well suited to buildings where space is limited, but a disadvantage is that the pens are next to the outer walls. Adequate heating and avoidance of draughts may therefore be difficult to achieve. This problem is overcome by having side passages (design 3, Fig. 7.4), but this pen configuration is less economical on space, since two feed passages are required.

For larger enterprises a suitable layout is the four-row design with central passages, combining economy of space with ease of access to the pens.

In winter, with pens adjacent to outer walls, it may be necessary to protect the kids from down draughts, by placing a solid top cover over the rear of the pen.

Floors and walls

Floors should be built of concrete over a damp-proof course, with straw bedding. The fall of the floor (1 in 20) should not be through other pens. The floor of the passage should have a non-slip surface. Walls should be easy to clean and free of projections, electricity cables and flaking paint. If possible the building should be accessible to mechanical cleaning.

1. Single row

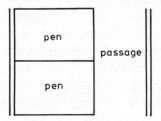

2. Two-row, central feed passage

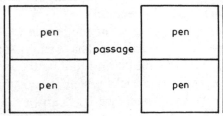

3. Double pens, side passages

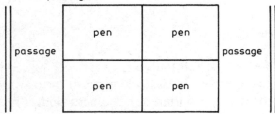

4. Four-row yards with central passage

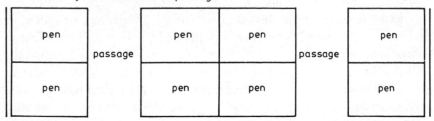

Fig. 7.4 Pen designs for kid-rearing.

Accommodation pre-weaning

After receiving colostrum, kids may be reared on milk substitute prior to weaning (see Chapter 8). Groups should be formed according to when the kids are born, allowing about 0.3 square metres of pen space per animal. The temperature in the pen should be kept at 14°C for the first 10 days. Subsequently the temperature may be reduced to 10°.

Milk substitute may be offered to the group in a trough or *ad libitum*, either cold or warm, via teats. The number of goats per teat may be varied from 2 to 20. If a milk dispensing machine is used, it should be located outside the pens and the pipes feeding the teats secured so that they cannot be chewed by the animals.

Creep feed should be provided in the pen in a trough positioned above ground level. The trough should have a hood or sloping front to prevent kids from jumping into it and fouling the feed.

Hay should be provided in racks which are positioned so that kids cannot jump into them – otherwise legs can be broken by animals getting them caught in the rack.

Clean water must be provided at all times, even when the animals are offered milk *ad libitum*.

Fields and fencing

Goats are often housed all the year, either because land for grazing is not available, or the fencing is inadequate, or the management of the pasture has been inappropriate for the herd and ill-health or poor performance has led to grazing being abandoned.

Shelter from rain and cold winds is useful, though not absolutely essential. Goats prefer access to shelter because they are less tolerant of adverse weather than sheep or cattle.

Fencing must be substantial – strong sheep netting with an electrified top wire 1.5 metres above ground level can be very effective provided the top wire is well maintained and electrified continually during the grazing season. If the fencing is not electrified, it must be strong enough to withstand the animals rearing on their hind legs and pushing with their front feet against the fence. Post and rail fences must have the rails set close enough to prevent younger stock from escaping.

Fields should be subdivided into paddocks with electrified netting or wires set at 0.3, 0.7 and 1.0 metre above ground level. A clean grazing system should be adopted if possible, to avoid accumulation of gastrointestinal worm eggs and larvae on the pasture. Clean grazing essentially involves resting a paddock or field so that the life cycles of the parasites are broken. Thus areas grazed one year are cut for silage or hay the following year.

The requirements for fencing vary with breed; Saanens are usually more easily restrained than other breeds. Animals reared with electric fencing learn to respect it, whilst newly-introduced adults require longer to adjust. Larger groups tend to be more easily restrained than animals which have been kept individually or in small herds.

Water must be provided to grazing goats at all times.

Milking installations and equipment

The various designs for milking installations range from portable bucket or churn units which are suitable for herds of up to 30 goats, to parlours suitable for milking large herds.

Portable units

The portable bucket or churn unit comprises a vacuum pump driven by an electric motor or petrol engine, two cluster assemblies, bucket and pulsation system, all mounted on a trolley. Clusters are available for either side or rear milking. The unit is economical, requires no fixed vacuum pipeline, may be used either indoors or outdoors, and is suited to farms where the herd may be kept in a number of different buildings and cannot be collected into one parlour for milking.

The disadvantages of the portable unit are that the number of animals which can be milked per hour is limited, cooling of the milk is slower than with a pipeline milking system in a parlour, and hygiene control is poor.

Parlours

A parlour is advisable for herds larger than 30 animals. It is important that the design and operation of the whole installation, including milk cooling and storage, should meet the hygiene standards set for dairy cows. The size of the parlour should fit the size of the herd so that all animals can be milked in a $1\frac{1}{2}$- to 2-hour period twice daily.

Collecting yards for parlours should allow 0.6 square metres per goat with steps up to and down from the parlour. Each entry step should be wide and shallow (e.g. four steps, each with 540 mm tread and 140 mm rise), whilst exit steps should be relatively deep with wider treads (e.g. two steps, each of 800 mm tread and 400 mm descent).

The design of the parlour must be considered very carefully so that it is correctly accommodated in the space available, adequate for the present and future size of the herd, and amenable to hygienic milk production. In particular there should be no direct contact between the bedded area and the parlour. All surfaces should be easily hosed down after milking.

Plate 7a A commercial goat milking parlour with 12 stalls and 12 milking clusters. (*Courtesy of Fullwood and Bland Ltd.*)

Most parlours are either of the abreast or herringbone type. The goats stand on a raised platform side by side either at right angles (abreast) or at 45 degrees (herringbone) to the area occupied by the operator.

Plate 7b Goats are often milked through their back legs. Note the identification tags. (*Courtesy of Fullwood and Bland Ltd.*)

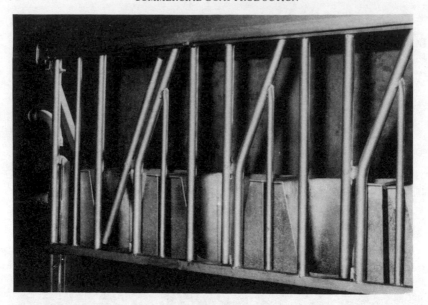

Plate 8 A self-locking cascade yoke system. The animal locks the yoke when its head enters the left-hand yoke. This triggers the next yoke to open and so on until all yokes are occupied. (*Courtesy of Food Research Institute, Shinfield.*)

The animals are held in position by yokes which may operate on a self-locking, cascade system. The first goat on the platform walks to the first yoke which is the only one open. The action of entering the yoke automatically locks the yoke and at the same time triggers the opening of the next yoke, and so on until all yokes are occupied. At the end of milking all the goats are released, and the yokes are ready for the next batch. The controls for moving animals in and out, and for delivery of feed, should be placed within easy reach of the milker.

The capacity of the parlour is determined by the total number of stalls, the number of cluster units and the number of persons doing the milking (see Table 7.4).

If milking is carried out by one person, single-sided, twelve-stall parlours can be operated with up to 150 animals. The maximum number of goats which can be milked is 90 per person per hour. However, it is advisable to adopt a full routine of udder cleaning pre-milking, and teat disinfection post-milking. This means that a more realistic rate of milking is about 50 animals per person per hour, allowing an average of 4 minutes from putting the milking cluster on to removal.

Table 7.4 Capacities of milking parlours

Number of sides	Total number of stalls	Number of milking clusters	Number of milkers	Maximum capacity (no. of goats)
1	6	2	1	40
1	12	4	1	80
1	12	6	1	100
1	12	12	2	150
2	24	12	2	200
2	24	24	2	300
2	36	18	2	350
2	36	36	4	500

The udder of the goat is more delicate than that of the cow. Therefore with machine milking a lower level of vacuum is used: the average is 37 kPa (11″ mercury) for goats compared to 50 kPa (15″ mercury) for cows. The rate of pulsation varies according to the individual manufacturer but is normally 70 to 80 pulsations per minute, with the milking phase comprising 50 per cent of the full pulsation cycle. Teat cup shells may be made of plastic or of stainless steel, with soft silicone teat cup liners.

Milking cluster units should include recording jars of 7 kg capacity so that individual milk yield recording is possible.

Other essential equipment, usually kept in a separate room from that containing the parlour, is the vacuum pump and motor, the pulsation regulator, a receiver jar (usually 20 to 25 kg capacity) and a pump to transfer milk from the recording jars to the receiver jar and to the bulk tank. Alternatively the milk may be piped directly to the milk tank, which is refrigerated.

Cooling

It is vital that goat milk is cooled as rapidly as possible to 4°C to slow down the rate of bacterial multiplication and prevent the production of taints caused by the hydrolysis of fatty acids. Refrigerated bulk tanks may be used for cooling and storing milk. At least 5 litres capacity per goat should be allowed.

Pasteurisation

Larger retail outlets for liquid milk usually stipulate that pasteurised milk is supplied by the producer. Pasteurisation involves heating to 63°C for 30 minutes or to 72.5°C for 15 seconds, to destroy bacteria and extend the shelf life of the milk to five days. Two types of pasteuriser are available – batch and continuous flow. Typically, batch pasteurisers are of small capacity (e.g. 225 litres), are electrically heated, and are fitted with a thermograph (temperature chart) and automatic controls.

Continuous flow pasteurisers, otherwise known as High Temperature Short Time (HTST) types, are of much higher capacity. The smallest model can handle 500 litres per hour. Their installation is only justified if they are used for 30 minutes or more per day. Thus they are economically worthwhile if the average daily production exceeds 250 litres.

In addition to the pasteuriser itself, it is necessary to have a cooler, a pump, a filter and a second bulk tank to hold the cooled, pasteurised milk prior to packaging.

Packaging and storing

If the milk is destined to be sold retail, then it may be packaged in cartons or in sealed polythene sachets each containing one pint, half a litre or one litre. For wholesale distribution, convenient package sizes are either 5 litres or one gallon.

Goat milk has the advantage that it can be frozen, though containers for freezing should not exceed 5 litres in size and the milk must be frozen immediately after being cooled, and stored at –20°.

Facilities and equipment for milk products

Many commercial goat producers add shelf life and value to their milk by processing it into cheese, yoghurt, cream or butter (see Chapter 9).

The essential facilities for making milk products are:

- a small room where starter cultures can be prepared
- a room for making cheese and yoghurt
- a room for storing and maturing cheeses

It is important to remember that the production of milk products is subject to statutory standards of hygiene and composition, and the premises are liable to inspection by Environmental Health Officers. If the products are sold from the farm, then the weighing and packaging equipment is subject to inspection by Trading Standards Inspectors under the Weights and Measures Act.

Cheeses and yoghurts require starter cultures of specific strains of lactic acid bacteria, to ensure the correct development of the product. Purchased cultures can be maintained by propagation in sterile milk under aseptic conditions. Thus it is important to keep starter cultures under strict conditions of hygiene, to reduce the risk of cross-infection by other bacteria or viruses, known as phages, which attack and kill bacteria.

Avoid entering the starter room without first changing overalls and scrubbing hands. Use a rotation of starters to avoid the build-up of phages against one particular strain of bacteria.

For cheesemaking the necessary equipment comprises a vat for producing the cheese curds, and moulds and muslin for making the cheeses. It is necessary to have some means of sterilising the equipment and of heating the milk to 70–72°C. Cheese rennet is also necessary to produce the curds.

For yoghurt production an incubator is required, in addition to equipment for heating the milk to 72°C. The temperature of incubation is normally 44–45°C.

For cream and butter production a mechanical separator is necessary. The very small size of the fat globules in goat milk results in slow, incomplete separation and risk of souring if the milk is left to stand. The separator requires fine and frequent adjustment for efficient cream production. The setting will vary according to the type of cream required (single or double), breed of goat, stage of lactation and season of the year. For butter the cream should be heat treated to 72°C, cooled to 45°C, and aged for 12 hours in a refrigerator before manufacture. A churn is necessary for producing the butter, which may be salted before the moisture is squeezed from the butter granules and the product is packaged.

Both goat cream and butter are naturally white unless a small amount of butter annatto is added to give the desired coloration.

Feed storage

If possible, all feeds should be stored under cover in dry, cool buildings, free of vermin. The overall size of the feed storage facilities will obviously vary with the number of animals in the herd, and the type of diet they receive (see Chapter 8).

Typical values for the bulk density and storage capacity of some common feeds are given in Table 7.5.

Silage may be stored in a walled bunker or an unwalled clamp, or in big bales sealed in polythene. The density of bunker and clamp silage varies with dry matter (DM) content, and to a lesser extent with length of chopping.

Table 7.5 Typical bulk density and storage capacity required for some common feeds

Feed	Typical bulk density (kg/m³)	Storage capacity (m³/tonne)
Clamp silage	660	1.5
Hay in conventional bales	145	6.9
Straw in conventional bales	82	12.2
Potatoes	640	1.6
Carrots	475	2.1
Brewers' grains:		
fresh	1000	1.0
ensiled	1300	0.77
Pressed sugar beet pulp	1400	0.71
Grains:		
wheat	785	1.3
barley	705	1.4
oats	513	1.9
peas	785	1.3
beans	833	1.2

Source: MAFF/ADAS (1981) *A Guide to the Estimation of Forage and Feed Stocks for the Winter Feeding of Livestock.*

As an approximate guide:

$$\text{density (kg/m}^3) = \frac{6500}{\%\text{DM}} + 400$$

for grass silage up to 35 per cent dry matter content. Thus the density of silage in a silo with a settled height of 2 to 3 metres may vary between 450 and 850 kg/m^3. The density of big bales is typically lower than that of clamp silage, at 300 to 400 kg/m^3. A bale of 1.2 metres diameter and 1.2 metres width weighs about 0.5 tonnes at 30 per cent dry matter content.

An approximate guide to the weight of big bale silage is:

$$\text{weight (kg)} = 727 - (7.0 \times \%\text{DM})$$

Big, round bales of straw typically weigh about 300 kg, but depending upon the size of the bale, the type of baler, and the type of straw, the weight may vary from 135 to 460 kg per bale.

Further reading

Alfa-Laval (1984) *Machine Milking of Dairy Goats*. Alfa-Laval Agri International AB, S-14700 Tumba, Sweden.

MAFF/ADAS (1984) *Goat Housing*. Design Data Package Booklet produced for 'Goat '84', 29/30 September 1984

MAFF/ADAS (1982–86) Various leaflets on livestock housing, farmhouse dairy products, milking parlours and equipment, milk hygiene and clean milk production.

Webb, K.R. (1985) *Buildings for the UK Goat Industry. A Basis for Design*, BSc(Hons) thesis, University of Reading, Early Gate, Reading, Berks.

8 Nutrition

For goats to be productive, they must receive an adequate supply of essential nutrients. The successful commercial goat producer therefore needs to understand:

- the digestive system of the animal
- nutrient requirements for specific levels of production
- factors affecting appetite or voluntary feed intake
- the composition and nutritive value of feeds

The basic principles of ration formulation are to provide adequate amounts of nutrients for the required level of production within the appetite capacity of the animal. In addition, this must be achieved as cost-effectively as possible.

Digestive system of the goat

Fortunately, the goat is a typical ruminant with a digestive system similar to that of cattle or sheep. Thus the principles of nutrition developed over many years with cattle and sheep apply also to goats.

Feed enters the stomach from the mouth, where it is mixed with saliva. The stomach of the ruminant is quite different to that of a human or a pig. The main difference is that there are four compartments: a large fermentation vessel called the *rumen*, and two smaller compartments – the *reticulum* and the *omasum* – through which the feed passes before it reaches the *abomasum*, or true stomach (Fig. 8.1).

The rumen contains a large population of micro-organisms, and digestion in the rumen can take several days, depending on the physical form of the feed, until the particles of feed are small enough to pass on down the digestive tract.

In young goats the rumen is undeveloped and during suckling milk is channelled through a groove (the oesophageal groove) into the true stomach (the abomasum). It is important that the reflex action of

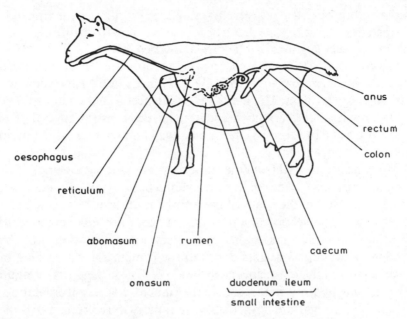

Fig. 8.1 Diagram of the digestive tract of the goat.

closing the groove occurs at all times when the kid is drinking milk, otherwise milk will enter the rumen and be fermented. This can cause bloat and digestive upsets. As more solid feed is consumed by the kid, the rumen enlarges and its microbial population increases. By the time the animal has reached adult size, the rumen comprises 85 per cent of total stomach capacity.

Digestion in the rumen

Whilst feed is in the rumen, it is continually being broken down both physically and chemically by the actions of the animal and the microbial population. The wall of the rumen is very muscular, contracting to mix the contents and to stimulate regurgitation for further chewing – the process of rumination.

The rumen microbial population of bacteria and protozoa secrete enzymes which digest the feed. The type and size of the population vary greatly depending on the diet, and changes should always be made gradually to allow adaptation by the microbes to new types of feed.

The residence time in the rumen depends on the speed with which feed particles can be broken down to a small enough size to allow them to pass on through the digestive tract. The exit orifice of the rumen is so small in sheep and goats that whole grains cannot pass through. For this reason whole grains can be digested efficiently.

Carbohydrates (fibre, starch, sugar) are fermented in the rumen to produce two main products, volatile fatty acids (VFA) and methane gas – a waste product. The main VFAs are acetic acid and propionic acid, though small amounts of butyric acid are also produced. The VFAs are absorbed through the rumen wall and are used as the main source of energy by the goat.

Methane is removed from the rumen by eructation (belching). This process must occur continually to prevent a build-up of gas and bloat, which can result in the death of the animal if immediate action is not taken. Two relatively common causes of bloat are sudden access to large amounts of concentrates, with very rapid digestion and the production of large amounts of gas in the rumen, and the feeding of fresh legumes such as clover or lucerne. These legumes contain proteins which form a stable froth in the rumen and prevent eructation.

Starch and sugar are fermented very rapidly in the rumen. All the sugar and about 90 per cent of the starch in the diet disappears in the rumen. Too much starch or sugar can cause digestive upsets such as acidosis, due to excess acid production in the rumen. By contrast, the fermentation of fibre is much slower, and therefore less likely to cause digestive upsets. When fibre is fermented, the main VFA produced is acetic acid, whereas the micro-organisms which ferment starches and sugars produce relatively large amounts of propionic acid. Apart from being used as sources of energy, the VFAs are also used in the synthesis of other important products. For example, acetic acid is used for synthesising milk fats. Glucose, used by the mammary gland to produce lactose, is synthesised from propionic acid. Therefore milk yield, which is governed by the rate of lactose production, is stimulated by starchy and sugary feeds, but milk fat levels may be depressed. Diets high in forage favour the production of milk with high levels of fat (Table 8.1).

Fibre stimulates saliva production which buffers the acids produced during digestion and helps to maintain an efficient fermentation in the rumen. Rations should therefore contain a high proportion of their energy in the form of digestible fibre. If possible, concentrates should be given in several daily feeds to maintain stable conditions for the rumen microbial population.

Proteins are broken down into smaller nitrogen-containing units – peptides, amino acids and ammonia, which the micro-organisms use to

Table 8.1 Typical percentages of volatile fatty acids in rumen fluid of sheep

Diet	Acetic	Propionic	Butyric	Others	Acetate: propionate ratio*
High fibre (e.g. 100% hay)	66	22	9	3	3:1
High carbohydrate (e.g. 60% concentrate, 40% hay)	52	34	12	3	1.5:1

* High ratio favours good milk fat levels

Source: McDonald, P., Edwards, R.A., and Greenhalgh, J.F.D. (1981) *Animal Nutrition* (3rd Edition) London: Longman

synthesise their own protein. This microbial protein is an important source of nutrients to the animal and is digested along with undegraded feed protein in the abomasum.

Ammonia is either used by the microbes to form microbial protein, or absorbed through the rumen wall and converted to urea in the liver. Some of this urea is recycled into the rumen via saliva.

The rumen microbial population needs adequate sources of nitrogen or protein to function efficiently. Therefore when dietary protein intake is low, recycled urea can be an important source of nitrogen to maintain rumen digestion. Some studies suggest that goats can recycle nitrogen very efficiently. Urea can be included in the diet as a source of supplementary non-protein nitrogen (NPN). But excessively high levels of ammonia in the blood, resulting from breakdown of urea in the rumen, are toxic to the animal. To reduce the risk of ammonia toxicity it is necessary to give NPN supplements in frequent feeds, preferably together with a source of readily fermentable energy, such as sugar or starch, so that the microbes can use the energy and ammonia from the NPN supplement rapidly to synthesise microbial protein.

The rumen microbes synthesise the B vitamins and vitamin K, and adult goats do not require supplementary sources of these vitamins. However, cobalt must be present in adequate amounts in the diet to meet the requirement for vitamin B_{12} synthesis.

Post-ruminal digestion

The rumen has a profound effect in modifying feed before it reaches the abomasum. Thus the material leaving the rumen contains:

- little starch and no sugar
- relatively little digestible fibre, relatively large amounts of indigestible fibre
- relatively little dietary protein; the nitrogen is mainly in microbial protein which has a fairly constant composition
- a high proportion of fat as microbial lipids, which are mainly saturated fats

Digestion of protein and fat by the animal starts in the abomasum and continues in the first part of the small intestine. Absorption of amino acids, fat and carbohydrates occurs in the small intestine, but there is further microbial digestion in the large intestine with absorption of the products. Absorption of minerals occurs in both the small and large intestines.

Feed intake

Goats are very agile feeders, with a strong preference for browsing rather than grazing. They appear to be able to tolerate a wider range of tastes than sheep or cattle and can make use of a wide range of feeds. They are also very selective eaters and have the advantage over other domesticated species of ruminants of being able to select a relatively high quality diet where a variety of feeds are available. However, they can be at a disadvantage when feed quality is uniform, particularly when it is uniformly poor. In this situation the animals may waste too much time attempting to select higher quality material, so that total intake is depressed.

There are also marked preferences for particular types of feed – especially if a choice is available. If possible it is a good idea to minimise the opportunity to discriminate against essential components of the diet such as high protein feeds, either by grinding or compounding a mixture into pellets. The animals will often reject mouldy or contaminated feed.

Most of the eating takes place during the day, with about 80 per cent of rumination occurring at night.

The factors influencing feed intake are:

- feed type and quality
- amount of feed on offer
- liveweight of the animal
- level of milk production
- frequency of feeding

With diets high in forage, intake is limited by the physical capacity of the digestive tract. Thus the amount consumed is determined by the size of the digestive tract, the rate of passage of feed through the tract and the space each unit of dry matter occupies in the tract. This third factor is a reflection of the digestibility of the feed, i.e. the amount which is absorbed from the tract as digested nutrients.

There is evidence that the rate of passage of feed particles through the gut is faster in goats than in sheep or cattle, but the size of the digestive tract appears to be a similar proportion of body weight in both sheep and goats.

Several studies have shown goats to have a higher intake of dry matter per unit of body weight than sheep, especially when given diets high in fibre. Selectivity, combined with a higher rate of passage, are important in achieving high feed intake.

Maximum dry matter intake in lactating goats is likely to vary between 4 and 7 per cent of body weight for individually-housed goats, depending on the type of diet and level of milk yield. Non-lactating adults are likely to have somewhat lower maximum dry matter intakes of between 2 and 3 per cent of lifeweight. Suggested maximum dry matter intakes for housed growing and lactating goats are given in Table 8.2.

Digestibility

Claims have been made that the goat is a more efficient digester of feeds than other ruminants, particularly when fibrous forages comprise the major dietary ingredient. But when the selectivity in feeding behaviour of the goat is eliminated (for example by grinding and pelleting the feed) goats may digest fibrous diets less well than sheep because of the characteristically higher rate of passage of food through the digestive tract. However, with higher quality feeds, differences between goats and sheep in digestibility may not be detected. In the absence of reliable data to the contrary, digestibility values for feeds derived from sheep or cattle may also be used for goats.

Table 8.2   Suggested maximum dry matter intakes for housed goats

Growing goats Liveweight (kg)	Dry matter intake (kg/day)
10	0.45
20	1.1
30	1.3
40	1.4

Adult goats	Milk yield (kg/day at 3.5% fat)						
	0	1	2	3	4	5	6
Liveweight (kg)	Dry matter intake (kg/day)						
50	1.5	1.7	1.9	2.1	2.3	2.4	2.5
60	1.8	2.0	2.2	2.4	2.6	2.8	3.0
70	2.1	2.3	2.5	2.7	2.9	3.1	3.3
80	2.4	2.6	2.8	3.1	3.4	3.7	4.0

Notes: Dry matter intake is assumed to increase with milk yield from 3% of body weight for
growing kids and dry animals to 5% of body weight for animals giving 6 kg milk/day.
In early lactation appetite is likely to be reduced

Nutrient requirements

Most of the information on the nutrient requirements of goats comes
from France and the USA. The data are limited and the recommenda-
tions in this section are a first effort to suggest requirements for the UK.
They are based on the first edition of *Nutrient Requirements for Goats*,
published by the US National Research Council in 1981, with
modifications based on French data, recent UK information from the
Animal and Grassland Research Institute, and dairy cow data.

Energy

Energy requirements are expressed as megajoules of metabolisable

Table 8.3 Suggested energy requirements of housed adult dairy goats

	Milk yield (kg/day at 3.5% fat)						
	0	1	2	3	4	5	6
Liveweight (kg)			Energy requirement (MJ ME/day)				
50	8.0	13.1	18.2	23.3	28.4	33.5	38.6
60	9.2	14.3	19.4	24.5	29.6	34.7	39.8
70	10.3	15.4	20.5	25.6	30.7	35.8	41.1
80	11.3	16.5	21.6	26.8	31.9	37.1	42.2

Notes: For grazing animals, increase daily energy requirement by 2.0, 2.3, 2.6 and 2.8 MJ ME for goats weighing 50, 60, 70 and 80 kg respectively
For milk of higher fat content, increase daily energy requirement by 0.6 MJ ME per percentage unit of fat

energy per day (MJ ME/day), and depend mainly on liveweight, amount of physical activity, weight change, stage of pregnancy, level of milk yield, and milk composition. Other factors influence energy requirements, such as diet composition and environmental temperature, but there are insufficient data to determine their significance for goats.

Table 8.4 Suggested energy requirements of housed young goats

	Liveweight gain (g/day)				
	0	50	100	150	200
Liveweight (kg)			Energy requirement (MJ ME/day)		
10	3.0	4.5	6.0	7.5	9.0
20	5.0	6.5	8.0	9.5	11.0
30	6.8	8.3	9.8	11.3	12.8
40	8.5	10.0	11.5	13.0	14.5
50	10.0	11.5	13.0	14.5	16.0
60	11.4	12.9	14.4	15.9	17.4

The energy requirements of housed adult dairy goats and of housed young goats are shown in Tables 8.3 and 8.4 respectively. They refer to animals which are relatively inactive. For grazing animals the requirements should be increased by 25 per cent. On extensive hill and upland grazing an increase of 50 per cent may be appropriate.

The values in Table 8.3 refer to animals producing milk of 3.5 per cent fat content and indicate a requirement of 5.1 MJ ME per kg milk, which should be adequate to meet a target minimum solids-not-fat content (SNF) of 8.0 per cent. But if milk fat or SNF content is higher, as is usually the case with Anglo-Nubians, for example, it would be wise to increase ME requirements by 0.6 MJ per percentage unit increase in fat and by 0.35 MJ per percentage unit increase in SNF.

Loss of weight in early lactation occurs when milk yield is at its peak and appetite is limited. This weight loss should be replaced in mid-lactation. Unfortunately the composition of weight lost in early lactation or gained subsequently is not known for the goat. From dairy cow information it is suggested that a liveweight loss of 0.1 kg will contribute 2.8 MJ of ME towards the total energy requirement, and 0.1 kg of liveweight gain requires 3.4 MJ of dietary ME.

Energy requirements for pregnancy take account of the fact that most of the growth of the foetus takes place in the two months immediately before birth. Therefore it is suggested that the goat requires an extra 6 MJ of ME per day during the last two months of pregnancy.

There is evidence that increasing energy intake during the last month of pregnancy by 7.4 MJ of ME per day improves subsequent lactation performance. Clearly, the level of nutrition in late lactation, i.e. in mid-pregnancy, will influence the level of the response. But the diet should be adjusted to avoid over-fat or over-thin animals at kidding, and careful attention should be paid to body condition throughout mid- and late pregnancy.

The energy requirements of housed young goats (Table 8.4) refer to loose housed animals of different liveweights and rates of growth. They include an allowance for the increased activity of kids compared to adults. There is evidence to indicate that kids may have a lower efficiency of use of ME for growth than lambs, and that high rates of daily gain are difficult to achieve under commercial conditions.

Protein

Proteins are the main constituents of the animal body and there is a continual need for protein for body tissue turnover and renewal, as well as for growth of new tissues and for milk production. In addition, efficient digestion of feeds in the rumen requires an adequate supply of nitrogen for microbial growth. The microbes may also require specific amino acids from dietary proteins.

Table 8.5 Suggested digestible crude protein (DCP) requirements of housed goats[1]

Growing goats	Liveweight gain (g/day)				
	0	50	100	150	200
Liveweight (kg)			DCP (g/day)		
10	35	45	55	65	75
20	46	56	66	76	86
30	50	60	70	80	90
40	53	63	73	83	93
50	61	71	81	91	101
60	69	79	89	99	109

Adult goats[2]	Milk yield (kg/day)						
	0	1	2	3	4	5	6
Liveweight (kg)							
50	51	106	161	216	271	326	381
60	59	114	169	224	279	334	389
70	66	121	176	231	286	341	396
80	73	128	183	238	293	348	403

(1) For grazing, increase daily DCP requirement by 25%
(2) Increase DCP requirement by 57 g/day in the last 2 months of pregnancy

COMMERCIAL GOAT PRODUCTION

In general, requirements for protein are related to those for energy. They are shown in Table 8.5, expressed as digestible crude protein (DCP). Unfortunately DCP takes no account of the need to ensure that highly productive animals, particularly young kids and females in early lactation, receive feed protein which is not degraded in the rumen. This additional requirement for undegraded dietary protein (UDP) is because microbial protein synthesised in the rumen is insufficient to meet the animal's total needs. The higher the intake of DCP from plant or animal sources (i.e. excluding NPN), the higher the intake of UDP, especially if the feeds were heat-treated during manufacture.

Table 8.6 Suggested requirements for calcium, phosphorus and magnesium

	Calcium (g/day)	Phosphorus (g/day)	Magnesium (g/day)
Maintenance:			
liveweight (kg)			
10	1	0.70	0.18
20	1	0.70	0.35
30	2	1.4	0.53
40	2	1.4	0.70
50	3	2.1	0.88
60	3	2.1	1.06
70	4	2.8	1.23
80	4	2.8	1.41
Plus additional require-ments for:			
liveweight gain (g/day)			
50	1	0.7	0.14
100	1	0.7	0.27
150	2	1.4	0.41
200	2	1.4	0.55
Late pregnancy	2	1.4	0.60
Milk (per kg)			
2.5 to 3.5% fat	2	1.4	1.0
3.5 to 5.0% fat	3	2.1	1.0

Minerals and vitamins

Suggested requirements for calcium, phosphorus and magnesium are given in Table 8.6, based on data derived from sheep and cattle. It is important to ensure the correct ratio of calcium to phosphorus in the diet. To avoid hypocalcaemia post-kidding, it is advisable to have the animal mobilising reserves of calcium from bone pre-kidding. Thus a diet relatively low in calcium is indicated for animals in late pregnancy.

Supplementary magnesium is necessary at all times, and the level of supplementation should be increased when the animals graze lush spring or autumn grass, otherwise hypomagnesaemia (grass staggers) may occur.

Potassium is plentiful ih feeds of plant origin and deficiencies are unlikely to occur. Similarly, if the diet contains adequate nitrogen as animal or vegetable proteins, then requirements for sulphur are likely to be met. But if urea or ammonia-treated straw is in the diet, supplementary sulphur should be included to maintain optimal rates for microbial protein synthesis. When NPN is in the diet, the amount of supplementary sulphur should be adjusted so that the ratio of total N to total S is 14:1.

Supplementary sodium and chlorine are both necessary in most feeding situations and are normally supplied as common salt. If added to a complete mixed diet salt should comprise 0.5 per cent of the total diet. Alternatively, salt may be added at about 1.5 per cent of the concentrate part of the diet.

Goats will normally consume salt readily and intakes in excess of requirements are not likely to be harmful. Salt may be used as a management aid; by placing blocks containing salt in less favoured grazing areas, the animals may be encouraged to move into them. The inclusion of low levels of salt in feed blocks encourages their consumption whilst high levels of salt act as a limitation to block consumption.

Trace elements are required by the goat, as for other ruminants, though requirements are not known precisely. It seems that the goat is more tolerant of copper than sheep are; thus the levels of copper normally found in cattle feeds do not cause problems. When levels of sulphur or molybdenum are high in the diet, copper availability is reduced and therefore the requirement for supplementary copper is increased.

Suggested dietary concentrations of trace elements, based on work with cattle and sheep, are given in Table 8.7. Since the concentrations

Table 8.7 Suggested dietary allowances for trace elements

Element	Concentration in diet (mg/kg dry matter)
Iron	40
Zinc	40
Manganese	40
Copper	10
Cobalt	0.11
Selenium	0.10
Iodine:	
housed	0.5
at pasture	0.15
with kale, cabbage or clover in diet, i.e. goitrogenic feeds	2.0

Table 8.8 Suggested requirements for vitamins A and D

	Vitamin A	Vitamin D
	(i.u. per day)	
Maintenance:		
liveweight (kg)		
10	400	84
20	700	144
30	900	195
40	1200	243
50	1400	285
60	1600	327
70	1800	369
80	2000	411
Plus additional requirements for:		
liveweight gain (g/day)		
50	300	54
100	500	108
150	800	162
200	1100	216
Late pregnancy	1100	213
Milk (per kg)	3800	760

of trace elements and vitamins vary greatly from feed to feed, they are normally included in mineral supplements and in compounded feeds at levels designed to avoid deficiencies. Occasionally herbage from particular areas may be particularly deficient in one or more trace elements – for example, some hill pastures are deficient in selenium – and deficiency symptoms are observed. Appropriate ways of rectifying these deficiencies should be devised in consultation with a nutritionist or a veterinary surgeon.

Vitamins A, D and E are the most likely to be deficient in the diet of goats. Suggested daily requirements of vitamins A and D for goats are given in Table 8.8. The suggested dietary concentration of vitamin E is 15 i.u. per kg dry matter.

Water requirements

It is vital that clean water should be available *ad libitum*, even when goats are grazing or consuming feeds with a high water content. Whilst goats are relatively resistant to short periods of deprivation, inadequate water intake over longer periods of time will lead to reduced feed intake and lower production.

Requirements for water vary with the environment, the type of feed on offer and the animal itself. Clearly, the warmer the climate and the lower the humidity the greater will be the requirement for water, though some tropical breeds of goat appear able to conserve water more efficiently than temperate breeds. Rainfall can reduce water requirements by increasing the surface moisture on grazed pasture. Humid conditions lead to lower grass dry matter content than hot, dry conditions.

The water content of different feeds will have a marked influence on water consumption by the animal. For example, one kilogram of silage at 25 per cent dry matter provides 750 grams of water, but one kilogram of hay at 85 per cent matter provides only 150 grams of water.

Since milk contains 87 per cent water, animals giving high milk yields have higher requirements for water than less productive animals.

Under temperate climatic conditions the requirement for water is about 4 kg water per kg dry matter consumed. At pasture, and on silage-based diets, water consumption will be low, but an animal eating 3 kg/day of a hay and concentrate diet may drink up to 13 kg water per day.

Composition of feeds

The main constituents of feeds are shown diagrammatically in Fig. 8.2.
Most feeds can be divided into two types – succulent feeds such as fresh
grass, silage, or roots, with dry matter contents in the range of 10 to 35
per cent, and dry feeds such as hay or cereal grains with dry matter
contents in the range 80 to 90 per cent.

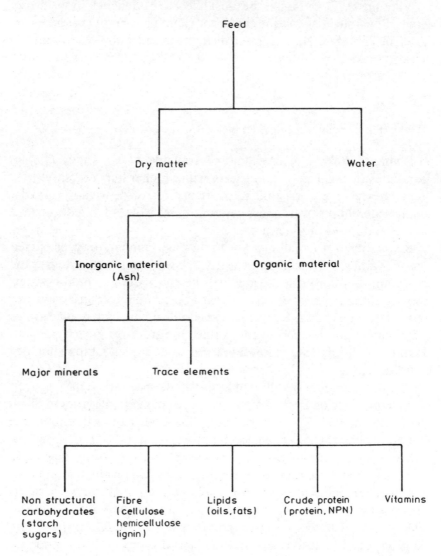

Fig. 8.2 Principal constituents of a feed.

The dry matter (DM) of feeds comprises organic and inorganic material (the 'ash' fraction). Most feeds contain between 5 and 10 per cent of the DM as ash. The organic fraction comprises:

- non-structural carbohydrates and lipids, which provide energy
- structural carbohydrates of plant cell walls, otherwise known as fibre, which provide energy and indigestible material
- crude protein, which provides the building blocks for animal tissues. Excess protein in the diet may be used as a source of energy
- vitamins, which have a variety of roles and which are essential in small amounts

Carbohydrates and fibre are the main components of feeds of plant origin. Cereal grains and root crops contain mainly non-structural carbohydrates whilst forages are relatively high in fibre. Oilseeds and the majority of feeds of animal origin contain large amounts of protein and lipids.

Non-structural carbohydrates are rapidly and almost completely fermented in the rumen, whilst fibre is only partially digested. The extent of fibre digestion depends on the amount of lignin present in the feed. Lignin, and other cell wall material bound to it, is resistant to microbial attack and passes through the digestive tract undigested. The other cell wall components, hemicellulose and cellulose, are potentially completely digestible, provided microbes can gain access to them before the feed is passed out of the digestive tract.

The proportion of fibre in a feed is a useful indicator of its probable energy value. Feeds low in fibre are good sources of energy whilst those high in fibre (such as straw) are not. But a few feeds such as sugar beet pulp are high in *digestible* fibre, and are good sources of energy. Thus it is important to assess the digestibility of the fibre fraction of a feed in addition to knowing the amount.

Lipids traditionally comprise the fraction which is extracted in ether (a solvent of fats) and which is termed the ether extract, although modern analytical methods can define the fraction more precisely. Digestibility of lipids is high and they are a concentrated source of energy. Lipids are therefore used to increase the energy concentration of a diet. The total diet should contain at least 2 per cent lipid to promote milk fat synthesis and a glossy coat condition in the animal. Most feeds contain more than 2 per cent lipid and oilseeds are a particularly good source. Whilst lipids are useful energy sources, the total level in the diet should not exceed 6 per cent of the DM, or digestion of fibre in the rumen may be depressed.

The crude protein fraction comprises all the nitrogen-containing material and is calculated as total $N \times 6.25$. It therefore contains true protein, and a variety of other non-protein nitrogenous (NPN) compounds such as amines, amides, and ammonia. Urea is also included in this fraction.

The typical composition of forages, concentrates and by-products are given in Tables 8.9, 8.10 and 8.11 respectively.

Table 8.9 Typical composition and nutritive value of forage and root crops

	DM (g/kg)	Crude protein (g/kg DM)	Crude fibre (g/kg DM)	ME (MJ/kg DM)	DCP (g/kg DM)
Young grass	150	265	130	12.1	225
Mature grass	250	116	288	9.0	80
Clamp silage:					
high quality grass	250	170	300	11.0	120
low quality grass	250	160	380	9.0	80
Lucerne	250	168	296	8.5	113
Maize	260	90	233	10.8	60
Big bale silage:					
high quality grass	400	170	300	11.0	120
low quality grass	400	160	380	9.0	80
Hay:					
high quality grass	850	132	291	10.5	90
low quality grass*	850	92	366	7.5	45
moderate quality					
lucerne*	850	225	302	8.2	166
good quality clover	850	161	287	8.9	103
Straw:					
good quality*	860	380	394	7.0	0
poor quality*	860	240	426	5.0	0
Good quality dried grass	900	187	213	10.6	136
Kale	130	137	200	12.1	122
Fodder beet*	180	63	53	12.5	50
Potatoes*	205	90	38	13.3	80
Carrots*	130	92	108	12.8	62

* Lipid content below 20 g/kg DM (2%)

Table 8.10 Typical composition and nutritive value of concentrate feeds

	DM (g/kg)	Crude protein (g/kg DM)	Crude fibre (g/kg DM)	ME (MJ/kg DM)	DCP (g/kg DM)
Grains/seeds:					
barley*	860	108	53	12.9	82
wheat*	860	124	26	13.5	105
maize	860	98	24	13.8	69
oats	860	109	121	12.0	84
spring field beans	860	314	80	12.8	248
peas*	860	262	63	13.4	225
whole linseed†	900	260	59	19.3	208
whole soyabean†	900	369	46	14.9	328
locust bean*	860	69	76	13.8	47
Extracted meals:					
rapeseed	900	413	104	10.9	343
soyabean*	900	503	58	12.3	453
palm kernel*	900	227	167	12.2	204
linseed	900	404	102	11.9	348
decorticated cotton cake	900	457	87	12.3	393
white fishmeal	900	701	0	12.5	195

* Lipid content below 20 g/kg DM (2%)
† Very high lipid content

Nutritive value of feeds

Rations are formulated to supply energy, protein, minerals and
vitamins to meet the estimated requirements of the animal. Therefore,
in addition to knowing the composition of feeds, it is also necessary to
estimate their nutritive value; that is, their ability to supply essential
nutrients.

In many countries the energy value of feeds is expressed as
metabolisable energy (ME). The ME value of a feed is a measure of the
energy which is available to the animal after deducting the energy in the
indigestible fraction, in urine and in methane loss.

Protein value is expressed either as crude protein (CP), or as

Table 8.11 Typical composition and nutritive value of by-product feeds

	DM (g/kg)	Crude protein (g/kg DM)	Crude fibre (g/kg DM)	ME (MJ/kg DM)	DCP (g/kg DM)
Maize gluten feed	900	262	39	12.5	195
Fresh brewers' grains	220	205	186	10.4	154
Ensiled brewers grains	280	205	189	10.4	154
Dried molassed sugar beet pulp*	860	99	203	12.5	80
Pressed sugar beet pulp*	180	106	206	12.3	60
Apple pomace (fresh)	230	60	184	8.0	30
Sugar cane molasses*	750	41	0	12.7	14
Middlings (wheatfeed)	880	176	86	11.9	129
Whey	66	106	0	14.5	91
Pot ale syrup	400	350	0	14.2	168
Meat and bone meal (medium protein)	900	527	0	7.9	411

digestible crude protein (DCP), though increasingly the degradability of the protein in the rumen is taken into account.

The ME and DCP values for commonly used feeds and by-products are given in Tables 8.9, 8.10 and 8.11.

In practice, nutritive value is often predicted from laboratory analysis. ME is predicted from fibre – usually determined as the fibre which is insoluble in acid detergent. Alternatively, ME may be predicted from a laboratory determination of digestibility.

When comparing the nutritive value of different feeds, both energy and protein contents should be expressed on a dry matter basis. For example, if barley has to be replaced in a ration by another feed, the first step is to select a feed with similar energy and protein contents in the dry matter, i.e. wheat, potatoes or sugar beet pulp. The second step is to check that requirements for energy and protein can continue to be met – if not, another feed may have to be included. Third, it is necessary to check that the new diet is likely to be eaten by the animal. Finally, the fresh weight of the new feed required to replace the equivalent amount of barley is calculated. An example calculation for ME is shown in Table 8.12.

Table 8.12 Example calculation of the weight of potatoes required to replace 1 kg of ME from barley

---

1. ME provided by 1 kg barley (*from Table 8.10*)

$$= \frac{\text{DM content of barley (g/kg)}}{1000} \times \text{ME of barley (MJ/kg DM)}$$

$$= 0.86 \times 12.9$$

$$= 11.09 \text{ MJ}$$

2. Fresh weight of potatoes required

$$= \frac{\text{ME from 1 kg barley fresh weight}}{\text{ME of potatoes (MJ/kg)DM} \times \left(\dfrac{\text{DM of potatoes (g/kg)}}{1000} \right)}$$

(from Table 8.9)

$$= \frac{11.09}{13.3 \times 0.205}$$

$$= 4.1$$

Therefore 4.1 kg potatoes would provide the same amount of ME as 1 kg barley

---

Monetary value of feeds

Having established that feeds can be substituted one for another on the basis of their nutritive value, it is also necessary to know whether it is worthwhile to purchase alternatives. The normal procedure is to calculate the unit value of ME and DCP for a standard mix and then to calculate the break-even monetary value of the alternative feed. For example, with barley at £110/tonne and soyabean meal at £150/tonne, the unit price of ME and DCP is £9.15 per MJ and £0.12 per gram, respectively (Table 8.13). On a nutritional basis alone, without taking into account delivery and handling costs, the break-even monetary value of brewers' grains as an alternative feed is £25/tonne.

Ration formulation

The principal steps in ration formulation are to:

1. *Calculate requirements* for energy and protein and to estimate likely DM intake (Tables 8.2, 8.3 and 8.5)
2. *Assess the nutritive value* of the feeds available. Forages and by-

Table 8.13 Example calculation of the unit value of ME and DCP for a standard mix of barley and soyabean meal, and the monetary value of fresh brewers' grains

---

(a) *Unit value of ME and DCP when barley costs £110/t and soyabean meal £150/t*

let x = unit value of ME (£/MJ)
y = unit value of DCP (£/g DCP)

On fresh weight basis nutritive value of barley:
11.1 MJ ME/kg; 70.5 g DCP/kg (*from Table 8.10*)

Nutritive value of soyabean meal:
11.1 MJ ME/kg; 407 g DCP/kg (*from Table 8.10*)

Solve simultaneous equations:
(i) $11.1 x + 70.5 y$ = price/t barley, and
$11.1 x + 407.7 y$ = price/t soyabean
(ii) $11.1 x + 70.5 y$ = 110, and
$11.1 x + 407 y$ = 150
(iii) x = £9.15/MJ ME
y = £0.12/g DCP

(b) *Monetary value of fresh brewers' grains*

Monetary value = $\dfrac{\text{% DM in brewers' grains}}{100}$

= [(unit value ME × ME content on DM basis) +
(unit value DCP × DCP content on DM basis)]
= 0.22 [(9.15 × 10.4) + (0.12 × 154)]

Monetary value of brewers' grains = £25/tonne

---

products should be analysed to determine their composition, since tables of typical feed composition may give inaccurate values

3. *Calculate the relative monetary value* of the feeds to ensure that they are worth using
4. *Combine the feeds* in appropriate amounts to meet nutrient requirements, taking into account appetite limits and other nutritional factors such as mineral and fibre contents

Examples of four different types of ration formulated to meet nutrient requirements for typical milk yields in early, mid- and late lactation are given in Table 8.14. The rations have been formulated assuming zero liveweight change in early and late lactation and a liveweight gain of 80 g/d in mid-lactation.

Table 8.14 Examples of rations formulated to meet the energy and protein requirements of a 70 kg lactating goat[1] giving milk of 3.8% fat content

		Stage of lactation		
		Early	Mid	Late
	Milk yield (kg/d)	5	3.5	2
	Ration (kg fresh weight/d)			
1.	*Early-cut silage[3]*			
	silage	5.7	7.6	8.0
	sugar beet pulp	0.6	0.5	0
	compound (12.5 MJ ME/kg DM; 14% CP)	1.4	0.6	0
	total DMI (kg/d)	3.2	2.9	2.0
	% forage in DM	45	67	100
2.	*Late-cut silage[3]*			
	silage	4.0	5.5	8.0
	sugar beet pulp	0.6	0.5	0
	compound (12.5 MJ ME/kg DM)	1.9 (16% CP)	1.2 (14% CP)	0.2 (16% CP)
	total DMI (kg/d)	3.2	2.8	2.2
	% forage in DM	32	48	92
3.	*Average quality hay[4]*			
	hay	1.2	1.25	2.25
	sugar beet pulp	0.6	0.5	0
	compound (12.5 MJ ME/kg DM)	2.0 (16% CP)	1.6 (16% CP)	0.45 (22% CP)
	total DMI (kg/d)	3.3	2.9	2.3
	% forage in DM	31	37	83
4.	*Straw and by-products*			
	straw[5]	0.6	0.8	1.0
	dried grass nuts[6]	0.6	0.5	0.3
	brewers' grains	1.5	1.3	1.3
	sugar beet pulp	0.6	0.5	0.2
	root waste[7]	5.5	5.0	2.0
	maize gluten feed	0.5	0.25	0.25
	total DMI (kg/d)	3.3	2.9	2.2
	% forage in DM	32	39	53

Footnotes on page 102.

(1) Milk yields are typical of those during early, mid- and late lactation for a goat giving 1000 kg milk in 305 days. Ration for mid-lactation (3.5 kg/d) assumes a liveweight gain of 0.08 kg/d. No allowance has been made for feed wastage
(2) Well-preserved silage 25% DM, 10.5 MJ ME/kg DM, 100 g DCP/kg DM
(3) Well-preserved silage 25% DM, 9.5 MJ ME/kg DM, 75 g DCP/kg DM
(4) Well-cured hay 8.5 MJ ME/kg DM, 45 g DCP/kg DM
(5) Bright clean barley straw 6.5 MJ ME/kg DM, zero DCP
(6) Good quality grass nuts, 10.6 MJ ME/kg DM, 187 g DCP/kg DM
(7) Mixture of carrots/potatoes etc., 17% DM, 13.0 MJ ME/kg DM, 70 g DCP/kg DM

With early-cut grass silage (Ration 1), a relatively low level of protein is required in the compound feed and this protein should be UDP rather than RDP. In late lactation sufficient silage should be eaten to meet requirements without the need for supplementary feeds.

The late-cut silage is assumed to contain less ME and DCP than the early-cut crop. Sugar beet pulp is a useful source of digestible fibre to maintain rumen function and milk fat levels in early lactation, when the proportion of forage in the total diet DM is relatively low. In late lactation a small quantity (70 g/day) of soyabean meal could replace the compound, since the silage meets energy but not protein requirements at the relatively low level of milk yield.

The average quality hay diet (Ration 3), like that for late-cut silage, requires considerable supplementation with both energy and protein in early and mid-lactation. Again, the inclusion of a source of digestible fibre is essential, especially in early lactation. In late lactation most of the energy requirements could be met by the hay, but protein supply would be very inadequate. As an alternative to a compound containing 22 per cent CP, it may be more economical to feed less of a 16 per cent CP compound with additional soyabean meal.

The ration based on straw (Ration 4) also includes dried grass nuts as a convenient source of supplementary energy and protein. Very good quality silage or hay may be used instead, but even so, the low quality of the straw and its low intake characteristics mean that the proportion of forage in the diet is relatively low. However, the fibre contained in the sugar beet pulp, and to a lesser extent in the brewers' grains and maize gluten feed, should ensure that reasonable milk fat levels are maintained.

Feeding strategies

Milking goats

The lactation curve of the goat is not well characterised – in some cases it appears to be similar to that of the cow, with a peak yield about four weeks post-kidding of about 0.5 per cent of total lactation yield, and a decline thereafter of 2.5 per cent per week. In other instances the production curve appears to be relatively flat. The steeper lactation curve may arise as a result of adequate feeding in late preganancy but under-feeding in early lactation.

High milk yields require a high nutrient intake by the animal in early lactation, whilst maintaining at least 40 per cent, and preferably 50 per cent, forage in the total diet. High-yielding goats tend to lose body weight in early lactation (80 to 100 g/day is a typical weight loss at this time), but their ability to maintain yield at the expense of body tissues appears to be less than that of the cow. For this reason the animal is likely to be more responsive to changes in nutrient supply than the cow.

Peak yield generally occurs before peak appetite, which may not be reached until 10 weeks post-kidding. Thus it is important to ensure that in early lactation the ration contains a sufficiently high concentration of nutrients, so that nutrient intake is adequate to support a high milk yield with a relatively small loss in body weight.

An important part of the overall feeding strategy is to identify, as far as possible, genetic potential for milk production and then to formulate rations which will supply adequate nutrients to meet that potential fully.

A herd average lactation yield of 1000 litres per head or more is a realistic commercial target. For Saanen-type goats the composition of the milk should average more than 3.5 per cent fat and more than 3.2 per cent protein. To achieve these targets economically it is necessary to have an ample supply of high quality forage so that supplementary feeds are only necessary in early and mid-lactation.

In practice the following strategy should be adopted to achieve the target lactation yield and the targets for milk composition:

- *Feed correctly in late pregnancy* to avoid over-thin or over-fat animals. Use high quality forage and introduce the lactation ration before kidding
- *Build up the concentrate part of the ration gradually* over a 4- to 6-week period post-kidding
- *Maximise forage intake* by:
  offering high quality forage
  offering fresh forage several times daily
  allowing at least 20 per cent wastage to ensure *ad libitum* consumption

- *Avoid giving more than 0.5 kg of concentrate* at a single feed
- *Provide variety* to stimulate appetite, but avoid the purchase of uneconomical feeds

The need for high quality forage applies to farms where silage or hay is home-produced. Here the cost of producing superior material may not be much greater than the cost of producing moderate quality silage or hay. But if all feeds are to be purchased, a suitable strategy may be to rely on poorer quality forages or straw, with greater reliance on by-product feeds.

Most dairy goats are fed on a step-rate feeding system where compound is offered to match the individual animal's actual milk yield. This approach may suit a mixed herd where high and low yielding animals are grouped together. A more logical approach is to divide the herd into groups according to stage of lactation and to give all the animals in the group a similar ration.

If high quality forage is in plentiful supply, it can be offered *ad libitum* with a flat rate of concentrate supplementation throughout the lactation. Differences in lactation yield between animals tend to reflect differences in forage intake, with higher yielding goats having a greater capacity for intake of silage or hay. Successful flat rate feeding depends not only on having high quality forage but also on making it available in excess of appetite at all times.

Pregnant goats

In early pregnancy the requirements of the foetus for nutrients are small; normally no additional nutrients are given until two months before kidding, when the animal is dried off.

Feeding in late pregnancy must be carefully controlled to avoid problems at kidding and to achieve maximal yields of high quality milk in the subsequent lactation. Some example diets are given in Table 8.15.

Both overfeeding and underfeeding in late pregnancy increase the risk of metabolic disorders in early lactation, particularly ketosis, and of reduced milk yields. Appetite can be depressed in over-fat goats, and body tissue reserves for milk synthesis in early lactation are minimal in animals which are thin at kidding. In addition, fat goats tend to have an increased incidence of difficult kiddings, whilst thin goats are more likely to give birth to weak kids.

Table 8.15   Example rations for late pregnancy

Ration[1] (kg/day)		Weeks before kidding	
		8 to 2	2 to 0
1.	*Early-cut grass silage*		
	silage	6.25	*ad libitum*
	compound (16% CP)	zero	gradually increase to 0.5 at kidding
2.	*Late-cut grass silage*		
	silage	7.0	*ad libitum*
	compound (16% CP)	zero	gradually increase to 0.6 at kidding
3.	*Average quality hay*		
	hay	1.75	*ad libitum*
	compound (22% CP)	0.35	gradually increase to 0.7 at kidding
4.	*Straw*		
	straw	1.0	*ad libitum*
	dried grass nuts	0.5	increase to 0.8
	brewers' grains	1.3	1.3
	root waste	1.5	increase to 2.0 at kidding

(1) 70 kg goat; for details of feeds see Table 8.14

Research in France suggests that if goats are given high-forage diets in late pregnancy then the animals continue to have relatively high intakes of DM in early lactation and to produce more milk than animals given diets low in forage. If possible, forage DM intake should be at least 17 g per kg of liveweight.

Work in the UK at the AGRI compared straw with hay for goats in late pregnancy. Intake of straw was low, and despite a higher concentrate allowance, total intake of energy was lower on the straw diet than on the hay diet. During lactation both groups were given the same daily concentrate allowance (1.7 kg/head) with *ad libitum* forage. The animals given hay in late pregnancy ate more forage in lactation and gave more milk than those given straw in pregnancy.

Supplementary minerals should be included in the diet in late pregnancy, but it is important to ensure that excess calcium intake is avoided to reduce the risk of milk fever. A 70 kg female requires about 6.0 g calcium and 4.2 g phosphorus per day during late pregnancy. Feeds high in calcium and low in phosphorus, such as kale and sugar beet pulp, should not be given at this time.

Males

Two feeding strategies should be adopted – one for the breeding season and one for the rest of the year, to take account of the fact that during the breeding season forage intake decreases markedly. Thus concentrates are normally given at 0.3 to 0.6 kg per day at this time. Higher levels may be given if the breeding season is short, with intense breeding activity. Concentrates should be introduced gradually about one month before the breeding season, commencing at about 0.1 kg/head/day.

For the remainder of the year the feeding level should be designed to maintain liveweight, or to rectify any loss in weight which may have occurred during the breeding season. Forage may comprise the entire ration but some variation between feeds is advisable.

Intake may vary from 13 to 16 g of forage DM per kg liveweight. Some supplementary feed may be necessary if forage quality is very poor, and intake may have to be restricted if forage quality is high.

Supplementary minerals and vitamins are required, particularly salt, but calcium (4 to 6.5 g/day) and phosphorus requirements (3.0 to 5.5 g/day) are lower for males than for milking females, as are requirements for magnesium. Excess intakes of phosphorus and magnesium may cause urinary calculi.

Herd replacements

The feeding strategy for rearing herd replacements should be aimed towards the production of well-grown, large framed females, mated to kid at 12 to 15 months of age. Animals kidding at 12 months should be 60 per cent of the normal adult weight when mated at 7 months of age, i.e. 42 kg liveweight for a 70 kg adult. The average daily weight gain to achieve this weight is 180 g – which is high, but achievable.

Animals kidding at 15 months of age should reach a target weight of

46 kg when mated at 10 months of age, with an average weight gain pre-mating of 140 g per day.

It is good practice to offer a variety of feeds during the rearing period so that the animal is adaptable to change in diet subsequently. It is also advisable to have a high forage diet to encourage forage intake not only during rearing but also in the subsequent lactation, to reduce feed costs.

Feeding after mating should follow the strategy outlined for the pregnant goat, taking into account the smaller body size of the goatling and the need to maintain growth of the animal during pregnancy.

During the first lactation it is necessary to supply energy and protein to give 8 to 10 kg of liveweight gain (about 30 g per day) in addition to that required for milk production.

Kids

Kids are usually allowed to suck their mothers for the first two days after birth to ensure they receive colostrum. It is important that adequate colostrum is consumed from the dam during the first few hours of life. If they do not, they should receive colostrum by bottle or by stomach tube.

Milk is often withheld from sale for the first five to six days after kidding; during this period kids can receive whole milk. Thereafter it is considerably more economical to rear kids on milk replacer. Calf milk replacer is adequate and the cheapest source of feed for kids, but there are many different formulations – some specifically for warm feeding and others for cold milk rearing. Thus it is good practice to find a source which fits the individual farm system.

Milk replacers should be given according to the manufacturer's instructions. The recommended concentration after mixing is between 120 and 160 g of powder per litre of mixture, although the concentration may be lower if the replacer is to be offered *ad libitum*. The usual composition of milk replacer is 26 per cent CP and at least 14 to 16 per cent fat. Whey from cheesemaking is not a suitable feed for young kids.

Kids can perform satisfactorily on *ad libitum* feeding systems, but the cost of the high consumption of milk replacer may not be justified if the kids are not weaned early.

The mineral and vitamin requirements of kids are not well-defined but appear to be similar to those of calves and lambs. In intensive kid-

rearing enterprises a multi-vitamin injection is often given as a
precaution against stress.

Suggested feeding regimes for weaning at 10 weeks, 8 weeks and 6
weeks are given in Table 8.16.

Table 8.16   Suggested feeding regimes for young kids

	Age at weaning (weeks)		
	10	8	6
Week	ml milk replacer[1] per feed (feeds/day)		
1 to 4	750 (3)*	750 (3)*	*ad libitum*
5	750 (3)	750 (3)	50% of volume in week 4
6	750 (3)	850 (2)	50% of volume in week 5
7	850 (2)	550 (2)	
8	850 (2)	550 (1)	
9	570 (2)		
10	570 (1)		

(1) Milk replacer contains 120 to 160 g milk powder/litre
* Numbers in brackets refer to number of feeds per day.

Milk replacer is considerably more expensive than solid feeds, thus
weaning should occur as early as possible – but at the same time
producing healthy, well-grown kids. Weaning may occur gradually or
abruptly. The key criterion is that the animals should be eating at least
100 g/day of solid feed at weaning.

Concentrate containing 18 per cent CP should be on offer *ad libitum*
from about two weeks of age; it may be a coarse mix or a small pellet.
To encourage consumption the concentrate should be palatable and
very fresh. Good quality hay should also be on offer.

Growth rates for milk-fed kids should be about 200 g per day. With
8-week weaning, up to about 13 kg of milk replacer and 5 kg of early
weaning concentrate will be required.

Grazing

There have been several studies on the behaviour and management of goats on range and hill pastures, but there is little information on the management of lowland pasture when grazed by goats. On extensively grazed range pastures the browsing and highly selective grazing habit of the goat allows the animal to obtain a diet of higher nutritive value than that achieved by cattle or sheep, which are less selective grazers.

Plates 9, 10, 11, 12 Hill pastures grazed by goats or sheep. Note the grazed seed-heads of the rushes, and the rejected clover in the goat pastures. (*Courtesy of Hill Farming Research Organisation, Edinburgh.*)

The grazing habits of the goat may be a disadvantage when the pasture is of uniform quality or when grazing is restricted; the goat may waste time attempting to select a higher quality diet. In addition, the faster rate of passage of feed through its digestive tract may be reflected in lower herbage digestibility than with sheep or cattle.

The preference of goats for aromatic herbage makes them useful animals for weed control. Thus they may be complementary to sheep on hill pastures. Studies in Australia, New Zealand and at the Hill Farming Research Organisation in Scotland indicate that goats may be particularly useful in maintaining re-seeded hill land. When areas of rush-infested re-seeded pasture and improved blanket bog were stocked with similar liveweights per hectare of either sheep or goats, the goats grazed the rushes severely compared to sheep, even when the supply of grass was plentiful. Heather was also grazed more extensively by goats than by sheep. A further important feature of these studies was that clover was eaten to a lesser degree by goats than by sheep, whilst the improved grass species were grazed more heavily by sheep than by goats. At the end of the season the areas grazed by goats contained a higher percentage of clover than those grazed by sheep.

Therefore, for optimum pasture utilisation it appears necessary to integrate goats with other ruminants. This would avoid excessive wastage due to the highly selective grazing habit of the goat. It would also avoid the depression in goat performance due to the animals being forced to consume a high proportion of the total herbage.

If it is necessary to graze goats intensively, they should be moved to different areas of land frequently and followed by another species of livestock.

Further reading

Agricultural Research Council (ARC) (1980) *The Nutrient Requirements of Ruminant Livestock*. Farnham Royal: Commonwealth Agricultural Bureaux.

Agricultural Research Council (ARC) (1984) *Report of the Protein Group of the Agricultural Research Council Working Party on the Nutrient Requirements of Ruminants*. Farnham Royal: Commonwealth Agricultural Bureaux.

Devendra, C. (1978) 'The digestive efficiency of goats' *World Review of Animal Production,* **14**: 9–22.

INRA (1978) *Alimentation des Ruminants*. INRA Publications, Versailles, pp449–467.

Interdepartmental Working Party (1983) *Mineral, Trace Element and Vitamin Allowances for Ruminant Livestock*. MAFF/ADAS.

MAFF/ADAS (1982–86) Various advisory leaflets on characteristics of feeds.

MAFF/ADAS (undated) *Feeding Dairy Goats*.

MAFF/WOAD (1984) *Nutrition and Feeding of Goats*. Leaflet prepared for ADAS exhibit, 'Goat '84,' NAC, Stoneleigh, Warwickshire.

McDonald, P., Edwards, R.A., and Greenhalgh, J.F.D. (1981) *Animal Nutrition* (3rd edition). London: Longman.

McCammon-Feldman, B., Van Soest, P.J., Horvath, P., and McDowell, R.E. (1981) *Feeding Strategy of the Goat*. New York: Department of Animal Science, Cornell University.

NRC (1981) *Nutrient Requirements of Goats*. Washington, USA: National Academy Press.

Russel, A.J.F., Maxwell, T.J., Bolton, G.R., Currie, D.C., and White, I.R. (1982) 'Preliminary studies on the use of goats in hill sheep grazing systems' *Animal Production,* **34**: 391

Throckmorton, J.C. (1981) 'The potential for goat production' in: Farrell, D.J. (Ed) *Recent Advances in Animal Nutrition in Australia*, pp173–178.

Volker, L., and Steinberg, W. (1983) *Vitamin Requirements of Goats – A Review*. Basle: F. Hoffmann La Roche & Co Ltd.

9 Profitable Milk Production

The single most important factor affecting the profitability of milk production is the average price received for each litre of milk or milk product sold. It is therefore vital that every litre of milk produced is of sufficient quality to be offered for sale. Second, it is clearly important that every litre of milk is actually sold, and that it is sold for the highest possible price.

Goat milk producers, in contrast to cow milk producers, do not have a guaranteed price for their liquid milk. This presents unique marketing opportunities to the goat producer:

- to seek as much revenue for dairy goat products as the market will bear
- to add value to products by attractive and individualistic processing and packaging

But the absence of a centralised marketing authority places the responsibility squarely on the shoulders of individual goat farmers to organise the marketing of their produce – if they fail to do it effectively and efficiently they will make little profit, if any at all. Thus it is imperative to understand the markets for dairy goat products and how to sell to them successfully, so that profitable outlets are established and maintained from year to year.

Markets

There are four types of market for dairy goat produce:

1. farm shops
2. retail shops
3. food distributors
4. milk processors

Farm shops

The farm shop is the traditional market for dairy goat produce. The producer sells direct to the consumer at the farm, or by local delivery. The advantages of the farm shop are that distribution costs are low, margins are high and the needs of the customer can be readily identified. The disadvantages are that the market may be limited and, if the farm is in a tourist area, seasonal. There is also a high labour commitment to manning the shop.

Clearly there are opportunities to develop the farm shop into a larger-scale business offering a wide range of produce, particularly if the farm is near a large centre of population or conveniently situated near a main road.

It is important to establish the regulatory position relating to the retail sale of food to ensure that the farm shop is operated within the law. For example, planning permission may be required for a shop.

It is well worth taking advice at the outset from the local Environmental Health Department to ensure that the shop and the produce sold through it meet the requirements of the Food Act 1984. The sale of food from a mobile food shop or a stall requires registration under the Act, and any food business must meet the hygiene requirements of the Food Hygiene (General) Regulations 1970. All equipment must be safe to operate and the premises themselves must provide a safe working environment for all people working in them. The provisions of the Health and Safety at Work Act 1974 apply to farm shops as to any other place of work.

Advice on operating a food business is usually free and willingly given. It is much better to comply with the law at the outset than to be prosecuted and convicted. Apart from being fined and ruining the business, human health is a very serious matter – a responsibility not taken lightly (see also page 118, milk quality).

Retail shops

The continued and steady increase in the number of shops selling dairy goat produce has led to a considerable expansion in the retail shop market. In addition to health food shops and delicatessens, super-markets now sell dairy goat products. The trend is toward increased emphasis on consistency of supply and high quality; in return retail food shops are very much larger outlets than farm shops, though some shops are in reality retail shops which carry many products from other farms.

The supermarket chains undoubtedly represent the greatest market opportunity to the commercial producer, but in addition to high standards of quality, the quantities required may be far greater than the individual producer can supply. Thus it may be more appropriate to work through an intermediary – a wholesale food distributor.

Food distributors

The wholesaler can introduce the producer to a larger market than that in the locality of the farm, but in return the margin to the producer is less.

Dairy produce is perishable, even though shelf life may have been lengthened by converting milk into cheese or yoghurt. Thus it is essential that the product reaches the retailer and customer well before it begins to deteriorate. Rapid, regular distribution is a vital part of selling goat products, particularly if the farm is in a remote area of the country. Wholesale food distributors can fulfil a key role in acting as local collection points for groups of food producers covering a wide range of different types of food. By using a food distributor, more economical use of time, packaging, transport and delivery can be achieved. Ultimately, the rapid sale through the wholesaler of a greater volume of produce in good condition can be of considerable benefit to the business.

Milk processors

The larger milk producer may not be able to commit time to processing, yet the market may not be sufficiently large to allow all the milk to be sold retail in the locality.

An alternative approach is to sell to a wholesale milk processor. There may be economic benefits over selling to a wholesale distributor, if a share of the value added to the milk by processing is passed back to the producer. But the *quid pro quo* is that the processor may require a contract whereby the producer enters into an agreement to supply a specified quantity of milk of a consistent quality.

Ideally, the processor should be prepared to accept frozen milk. This allows the producer to store excess output and to maintain a regular supply during periods of relatively low milk production.

Marketing

Marketing is essentially the process of bringing products to the customer. The choice lies in whether to market directly to the customer or to employ a marketing agent. As with wholesalers, marketing agents are not charitable organisations and they demand a share of the margin.

It is generally preferable for the producer to market the produce directly, if possible. Not only is a higher margin achievable, the customer's interest in the product is more likely to be aroused by the direct contact with the producer than with an intermediary.

A strategy is necessary in marketing dairy goat produce so that the product is sold, distributed and supported correctly, to maintain outlets.

Priorities need early identification. For example – how is the product to be priced? Will there be seasonal price variations, or bulk discounts? How much money should be committed to promotion?

Other issues which require consideration in developing a marketing strategy are these:

- What types and sizes of product are required by different customers?
- How is each different product to be described and priced?
- How is each product to be promoted, distributed and supported?

Selling

The identification of prospective customers is the key to successful selling. Having found the interested customer, closing the sale is the next priority.

Having established the amount and the price, and achieved the sale, the next important issues are to arrange for delivery and to receive payment. The terms of business should be clearly understood by both parties. The seller should make it obvious that payment is expected within a specified period of time after delivery of the goods to the customer.

Having made an initial sale, it is essential to follow up the contact to establish customer reaction. If the product is not well received, there may be good reasons for the producer to change the specifications of the product, thereby increasing the probability of establishing regular customers and customer loyalty to the product.

Pricing is clearly a very important element in marketing and in selling. It is always better to set a high price, and if necessary to negotiate discounts for quantity or for rapid payment; it is considerably more difficult to raise the price having once established it at an inadequate level.

Demand

For many years dairy goat products were consumed mainly by people with allergies to cow milk, especially babies. Estimates have been made of the extent to which there is a genuine health-food element in goat milk. In one survey, 7.5 per cent of infants showed allergic symptoms to cow milk. Of these, about two-thirds were probably not allergic to goat milk.

Recently there has been a strong trend towards the consumption of dairy goat products by people who demand a greater variety of foods and who like the flavour of goat produce. This development has been led by the importation of goat products from France, Greece, Italy and other European countries. But it is also probably linked to greater mobility of consumers, and a wider variety of foods offered for sale in supermarkets and delicatessens. A further factor is that dairy goat products have an image of being 'natural' products. With the current prevalent concern over food additives, a 'natural' image is clearly an advantage.

Prices

Increased demand and increased consumption have occurred despite relatively high prices for goat compared to cow milk products, suggesting that consumers are prepared to pay substantial price premiums for products which they consider to be of superior quality. Some typical prices for dairy goat products in the UK are given in Table 9.1. They are not definitive and should be used as an approximate guide only. For budgetary purposes it is wise to negotiate up-to-date prices with potential customers.

The prices quoted in Table 9.1 may reflect the fact that the marketing of dairy goat produce has been aimed at higher income groups, and at those people who are willing to pay higher prices for speciality products and 'health' foods.

Table 9.1   Typical UK prices for goat milk and milk products, 1986

	Wholesale	Retail
Whole milk (p/litre)	30 to 45	60 to 70
Cheese (£/kg):		
hard	5.50 to 7.50	6.50 to 10.0
soft	4.50 to 5.50	5.50 to 7.50
Yoghurt (£/kg):		
plain	1.10 to 1.30	1.50 to 1.70
with fruit	1.25 to 1.40	1.70 to 2.00

Prices are subject to large seasonal variations, reflecting over-production in spring and summer following kidding, and under-production in autumn and winter. For example, recorded herds in France showed a very marked seasonal kidding pattern in 1983, with almost 80 per cent of total kiddings in January, February and March (Fig. 9.1). The price paid to producers varies inversely with the quantity produced, in an attempt to encourage winter milk production. Thus in 1984 the price paid to French producers in January was 24 per cent higher than that paid in July.

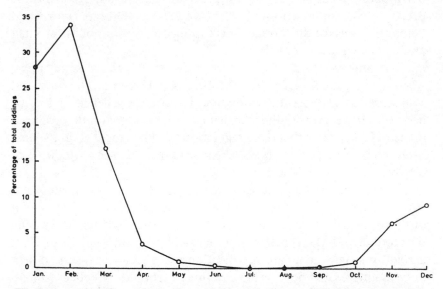

Fig. 9.1 Annual kidding pattern of recorded herds in France, 1983. Source: Scott, S.A. 1985 *Report on commercial goat farming in France.* MAFF/ADAS p.7.

Milk quality

The law

The sale of goat milk, cheese and yoghurt in the UK is subject to several regulatory controls, specifically:

- the Food Act 1984
- the Weights and Measures Acts 1963 to 1979
- the Trade Descriptions Act 1968
- the Food Hygiene Regulations 1966 and 1970

The principal regulations are those of the Food Act 1984 which make it an offence for any person to prepare or sell any food which is injurious to health. It is also an offence to sell any food which is not of the nature (e.g. chalk not cheese), substance (e.g. not cheddar but brie), or the quality demanded. In other words, any food which is not correctly described, or is of substandard quality, may render the seller liable to prosecution under the Food Act.

Authorised officers, usually Environmental Health Officers, are empowered under the Food Act to examine food offered for sale and to seize unfit food.

The Food Act also contains controls in relation to food labelling, and these may be increased in 1987 to be more specific in relation to nutritional labelling. At the time of writing (1986) they include declaration of ingredients and added water, durability marking ('best before'), claims for the product, misleading descriptions of the product and its presentation for sale.

Goat milk is not covered by the Milk and Dairies Regulations, which set standards of hygiene for milk production on the farm and standards for milk pasteurisation, and require the seller of milk to be licensed. However, with the increasing amount of goat milk produced in the UK, it is likely that these regulations will be amended in the near future to include both goat and sheep milk production in addition to cow milk.

The Weights and Measures Act and the Trade Descriptions Act cover the description of milk and milk products when sold by weight or by volume. It is important to understand the law when describing the product and when packaging it for sale. For example, if the product is sold by weight, there are specifications for the accuracy and design of weighing equipment. The equipment should be checked and stamped with the inspector's seal to ensure that it complies with the law.

The Food Hygiene Regulations cover the standards of hygiene for catering establishments, including people who prepare food at home. They also cover stalls at markets or exhibitions and food delivery vehicles. Environmental Health Officers inspect premises, stalls, and vehicles, to ensure that hygiene is adequate.

Compositional quality

Targets for the compositional quality of goat milk are given in Table 9.2. Unlike cow milk, there is no direct financial reward for the production of milk with a relatively high content of total solids, fat or protein. Thus the producer who confines his enterprise to the production of milk rather than cheese is unlikely to suffer financially from poor compositional quality milk unless it is linked to low yields, or the milk fails to meet minimum standards set by a processor.

But, for cheese and yoghurt production, the higher the compositional quality of the milk the higher the yield of cheese and the better the quality of the yoghurt. The targets in Table 9.2 should be regarded as minimum levels for milk destined for processing.

The composition of goat milk varies with type of diet (see Chapter 8) and with breed. Many commercial producers deliberately keep

Table 9.2 Targets for milk quality

Compositional quality		
(Saanen-type, % of fresh weight)		
total solids	>11.6	
fat	3.8 (≮ 3.5)	
protein	3.4 (≮ 3.0)	
lactose	4.3	
ash	0.8	
Hygienic quality		
antibiotics	absent	
taints	absent	
bacteria:	*Total bacteria*	*Coliforms*
milk sampled direct from udder	<100/ml	absent in 1 ml
freshly packaged milk	<1000/ml	absent in 0.1 ml

Note: milk with a total bacterial count of > 50 000 should not be sold to the public

animals of more than one breed, or crossbreds, in an attempt to avoid low levels of milk solids. In particular, Anglo-Nubians can boost compositional quality because their milk characteristically contains higher levels of fat and protein than that of other breeds (Table 9.3).

Table 9.3   Average milk yield and composition of British goat breeds

Breed (no. recorded)	Yield[1] (kg)	Composition (%)	
		Fat	Protein
Crossbred (229)	1220	3.7	2.8
British Saanen/Saanen (197)	1243	3.7	2.8
British Toggenburg/Toggenburg (97)	1169	3.7	2.7
Anglo-Nubian (75)	1040	5.0	3.5
British Alpine (73)	1099	4.1	3.0
Golden/English Guernsey (13)	992	4.1	2.9

(1) All lactations were of 270 days or more. Most were recorded for 365 days
Source: British Goat Society (1985) *Yearbook*, pp 35–45

Milk from freshly-kidded goats (colostrum) should be withheld from sale or processing for at least the first three days after kidding. Apart from possible taints from colostrum, it also contains antibodies which the new-born kid should receive to acquire immunity to infectious diseases. Colostrum should therefore be given *ad libitum* to kids for the first two to three days of their lives.

Tainted milk is a more common problem in goat than in cow milk production. It can be the result of using feeds which have specific components which are secreted in milk and which cause off-flavours (see Chapter 8). More often tainting results from lipolysis following the agitation of warm milk, or from lipolysis during slow cooling and inadequate refrigeration during storage. A further cause of tainting is a high bacterial count in the milk.

Milk should be cooled to less than 10°C within 30 minutes of milking and either stored under refrigeration at 4°C, or frozen immediately to –20°C, and stored at –20°C.

Where tainted milk is a problem, first identify the source of the taint. Fresh milk should be collected without agitation and sampled as soon as possible after milking. If taint is detected in the fresh sample then it most likely has its origins in the feed, or is due to an abnormal condition in the animals, such as mastitis or ketosis. If no taint is detected in fresh milk then the problem probably lies in lipolysis, and the processes following milking should be examined to ensure that either agitation is eliminated or the speed of cooling increased. Milk can be examined in the laboratory to determine if the problem is a microbiological taint.

Hygienic quality

Goat producers are at risk of prosecution if they sell milk which is unhygienic. An outbreak of salmonellosis in babies traced to goat milk would not only destroy the business of the individual producer, it would put the industry's reputation in jeopardy and may force other, innocent, producers out of business.

Salmonellosis and other diseases of humans known collectively as 'food poisoning' are on the increase. For this reason alone it is vital that producers adopt good hygienic practices so that their milk is not only free of pathogenic organisms but also has a relatively long shelf life.

Pasteurisation is the obvious way of reducing the risk of producing unhygienic milk. Hopefully, efficient small-scale pasteurisation equipment will become available in the near future for the smaller producer. In the meantime it is crucial that targets for the hygienic quality of unpasteurised milk are set and reached. These targets are given in Table 9.2. Surveys have shown that problems with goat milk are more likely to be the result of poor hygiene during and after milking than due to the transmission of bacteria from the goat herself. The most common problem is high counts of non-pathogenic bacteria (see Table 6.3, page 45), though if milk is not kept chilled it is a good medium for the growth of pathogens.

Milk taken directly from uninfected udders should contain not more than 100 bacteria per ml, with coliforms absent in 1 ml. Where good sterilisation of equipment is practised, freshly packaged milk should contain less than 1000 bacteria per ml, with coliforms absent in 0.1 ml (see Table 9.2). Unpasteurised milk with a total bacterial count of more than 50 000 per ml should not be sold to the public.

If a shelf life of several days is required, then rapid cooling and refrigerated storage is vital, even with clean milk. In one trial, milk with an initial count of 800 bacteria/ml was stored at 4°C, at 10°, and at a moderate room temperature. The refrigerated milk kept at 4°C was still of good quality, with a count of only 25 000 bactera/ml after four

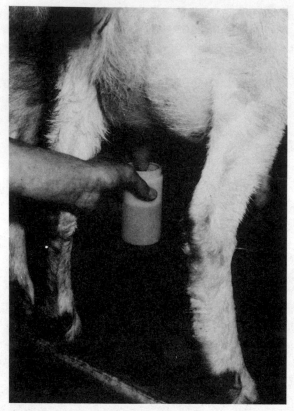

Plate 13 Teat dipping with disinfectant after milking helps to prevent mastitis and other infections of the udder. (*Courtesy of Food Research Institute, Shinfield.*)

days' storage, whilst the count exceeded 200 000 bacteria/ml after two days' storage at 10°C and after only 1 day's storage at room temperature.

The important management factors to bear in mind in order to achieve the targets for hygienic milk production are:

- ensure the goats are free from udder infections
- ensure the udders are clean at milking
- sterilise all milking equipment thoroughly and regularly
- cool milk as rapidly as possible, but without agitation
- store milk at 4°C or freeze until used or sold

The Department of Agriculture and Fisheries for Scotland has published an excellent code of practice on the hygienic control of goat milk. The code covers the design, construction, maintenance and operation of housing, milking areas and milk rooms, the sterilisation of milking equipment and hygiene at the time of milking. In addition, guidelines are given on the filtration, cooling, freezing, packaging and transportation of milk.

Outbreaks of disease involve isolating the affected animals so that they can receive suitable therapy. If antibiotics are used as part of the prevention or treatment of disease, the drugs will be excreted in milk. Contamination of milk can occur when the drugs are given orally, or by injection, or by insertion into the teat. The drugs can interfere with cheese and yoghurt production and can harm consumers by causing allergic reactions or build-up of resistance to the antibiotic. Therefore it is important to ensure that all milk which is sold for liquid consumption or used for processing is free of antibiotic residues. Veterinary advice should be sought on the appropriate withdrawal periods for specific drugs.

Output

The income received by the producer is determined by the quantity of product produced, its quality and the price at which it is sold. All three factors are clearly of fundamental importance to the success of the business, but if the required output cannot be achieved then customers will be lost and the growth of the enterprise will be stunted.

Milk and cheese

The lactation yield of the animal (Fig. 9.2) is the principal output from the dairy herd, and is influenced by a number of factors including breed, nutrient intake, health status, and the age of the individuals in the herd.

Breed differences in average lactation yields are relatively small, but the averages in Table 9.3 hide huge ranges in output, and in composition, within each breed type. Thus the range in yield within the Saanen-type was from 100 to 2200 kg of milk.

Since there is a wide within-breed range in performance it is likely to be as profitable to select within a breed for increased output as to change breeds.

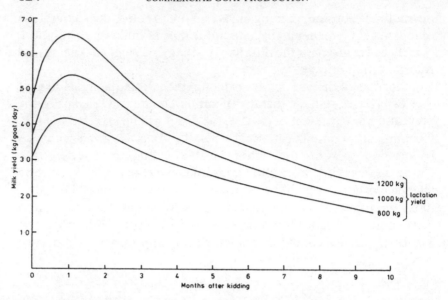

Fig. 9.2 Lactation curves of goats giving 800, 1000 and 1200 kg milk in a 10 month lactation. Source: Dudouet, E. (1982) Courbe de lactation theoretique de la chevre et applications. *Le Point Veterinaire* **14**: 53–61.

Nutrient intake should match the potential of the animal for milk output. Potential yield is likely to be reached when:

- nutrition pre-kidding is geared to achieving high milk yields in early lactation
- females are neither too fat nor too thin at kidding
- weight loss in early lactation, and weight gain in mid-lactation, are not excessive
- milk fat and milk protein contents are at target levels

Age of animal affects output, particularly in the first lactation. Goatlings kidding at two years of age generally yield about 70 per cent of the yield of adult milkers. Peak yields occur between the third and fifth lactations; thereafter output tends to decline.

Typical yields of cheese are given in Table 3.3, page 16. In general, the yield of soft cheese is likely to be about 25 per cent higher than that of hard cheese. Yields of both soft and hard cheeses are influenced by the compositional quality of milk; the higher the content of protein and fat, the higher the yields of cheese. Milk should contain at least the target values for compositional quality shown in Table 9.2 to achieve acceptable cheese yield and quality.

Seasonality of output

The seasonal breeding cycle of the goat results in the majority of naturally-bred adults being in early lactation in late spring. The consequence is that production virtually ceases in February and March, to be followed by a rapid rise after kidding (Fig. 9.3).

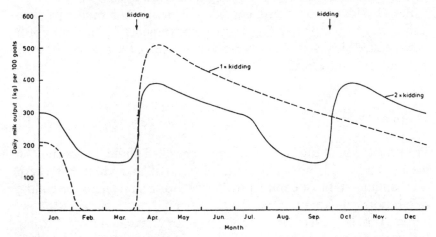

Fig. 9.3 Pattern of milk output from a herd kidding either as one group in spring or as two separate groups in spring and autumn.

Many customers require a constant supply throughout the year, and out-of-season kidding can overcome the large seasonal fluctuations in output. In Fig. 9.3 the total output each month from a 100-head herd is illustrated either as a naturally-bred herd or as a split herd with half kidding in April and half kidding in October. With the split herd there is a relatively low level of output in February and March, and in August and September. Peak output is more than 100 kg lower than that of the naturally-bred herd, and there is no time when the output of the herd is likely to be zero.

Kids

The number of kids born per goat can vary from one to three, to give herd kidding percentages in the range 150 to 220 kids per 100 goats kidding.

The value of kids sold depends on the number born, the proportion of males to females, the number of females required to be reared as herd replacements, and the age at which the kids are sold. For example, male kids may be destroyed at birth because it is not considered economically worthwhile to rear them for meat. By contrast, where markets for meat have been developed, kids may be worth between £5 and £8 at 10 days of age, and between £12 and £17 after weaning at 8 weeks of age, depending on their weight. The typical price paid by meat producers for kids is £1.00 to £1.20 per kg liveweight.

Female kids may be worth considerably more than males, particularly if they are sold for rearing as herd replacements. Typical prices for older female kids, weaned at more than 8 weeks, for rearing for milk production, are from £30 upwards.

Culled females

The proportion of the herd which may be culled annually depends on the age, milk yield and health status of the herd, and whether the herd is expanding, static or contracting in size. For a static herd the culling rate is normally about 20 per cent, compared to 25 per cent in cow herds. This is partly a reflection of a lower incidence of chronic infections such as mastitis, and partly a reflection of the fact that goats tend to remain productive for many years. This may in turn be the result of good management of individuals in the herd, particularly high yielders.

A further factor which may contribute to a low culling rate is the low value of the culled animal. The typical price range is £20 to £30 per live animal.

Variable costs

Variable costs are those costs which vary with the size of the enterprise. By far the largest variable cost is feed, which accounts for 75 to 85 per cent of total variable costs. Within the overall cost of feed, concentrates can comprise a low or a relatively high proportion, depending on the feeding strategy. Thus the provision of high quality silage can reduce concentrate costs whilst the lower energy and protein content of moderate quality hay is reflected in greater concentrate use and therefore greater feed cost (Table 9.4). By-products – if available – can substitute for both concentrates and conserved grass, with savings in total feed costs.

Table 9.4 Annual feed budgets per goat

	Diet[1]							
	Silage				Hay		By-products	
Forage quality	High		Low		Moderate			
Feed[2]	(kg)	(£)	(kg)	(£)	(kg)	(£)	(kg)	(£)
Silage	3330	60	2900	52				
Hay					720	36		
Sugar beet pulp	103	11	108	12	108	12	139	15
Compound	180	26	320	46	448	66		
Straw							430	11
Dried grass nuts							186	17
Brewers' grains							550	14
Root crop waste							1560	28
Maize gluten feed							75	9
Minerals		2		2		1		4
Total cost (£)		**99**		**112**		**115**		**98**

(1) From Table 8.14 plus wastage: silage and hay 25%; straw 40%; brewers' grains and root crop waste 10%; other feeds 5%

(2) Costs per tonne fresh weight: silage £18; hay £50; straw £25; compound 14% CP £143; compound 22% CP £156; sugar beet pulp £110; dried grass nuts £90; brewers' grains £25; root crop waste £18; maize gluten feed £115

Feed budgeting

Feed budgeting is an important management aid to profitable milk production. The exercise involves assessing the quality of bulk feeds and the nutrient requirements of the herd for a specified level of milk output, then calculating the quantities of bulk and concentrate feeds needed to meet the requirements and their costs. The data in Table 9.4, derived from the rations shown in Table 8.14, illustrate that although silage is more expensive per tonne of dry matter than moderate quality hay, the saving in concentrates more than compensates. This is because the energy and protein contents of both silages are higher than those of the hay – a feature consistently found on farms where silage making has replaced haymaking.

Other variable costs

Other variable costs include the herd health scheme (see Chapter 6) which has proved invaluable in larger herds, where prevention of infectious diseases is particularly important. In cheese or yoghurt production, the cost of starter cultures, cartons and packaging materials can amount to a significant item of expenditure.

(Typical variable costs for a 100-head herd are shown in Table 9.6.)

Fixed costs

Establishment costs

The cost of establishing a dairy herd depends on the number and type of stock. If the decision is made to purchase kids and to rear them, then clearly capital costs will be lower than if yearlings or adults are bought. The advantage of purchasing kids is that the herd is stable when kidding commences and there is less risk of individual animals not adapting to being housed in larger groups.

Reducing stress on the individual helps to reduce the incidence of ill health and leads to higher milk production. If goatlings or adults are to be purchased, they should therefore be given as long a time as possible to adapt to the environment and to their herd mates before kidding and milking commence.

Typical costs for the purchase of goatlings at one year of age, and adults, are: £80 to £120 for goatlings and £120 to £160 for adults.

Capital items

Some typical costs of milking and dairy equipment are shown in Table 9.5. It is advisable, whatever the size of the herd, to install the milking equipment in an area separate from that in which the animals are housed and fed. A further separate room for the dairy equipment should also be available, which can be kept sterile. Sterility in the dairy is particularly important if milk is to be sold to the public without pasteurisation, or if the milk is to be processed on the farm into cheese or yoghurt (see Chapter 7).

Other capital items necessary for the herd include:

- feeding and handling facilities
- artificial lighting (for out-of-season breeding)
- machinery for handling feeds
- delivery vehicle

The total capital which might be required to establish a herd of 100 goats is likely to amount to between £15 000 and £25 000, depending on the individual farm and the products to be sold from the herd.

Table 9.5 Typical costs of milking and dairy equipment, 1986

	Typical cost (£)
Milking parlours, complete with cooling equipment:	
herringbone, up to 150 goats 12/12	8 000
herringbone, up to 300 goats 24/24	12 000
Milking equipment:	
self-locking yokes and feed trough (6 goats)	550
single bucket unit, 20 litre capacity + 1 cluster	200
single bucket unit, 20 litre capacity + 2 clusters	250
portable milking unit, 2 buckets, 4 clusters	750
churn, 15 litre capacity	75
Dairy equipment:	
cooler, in churn	60
cooler, pump, filter	750
standby generator	750
pasteuriser, 18 litre capacity	350
pasteuriser, 250 litre capacity	2 200
bulk tank	1 500
cartons, per 100	4 to 10
cartoning machine, hand operated	5 000
cartoning machine, automatic	11 000
Cheesemaking equipment (*vat, press, moulds, etc.*)	500
Incubator for yoghurt (capacity 140 pots)	550
Freezer, domestic chest, 0.4 m³ capacity	200

Labour

The largest single recurrent fixed cost is labour, which is very often not fully taken into account when assessing the profitability of the enterprise. Since the cost of employing a person to milk the animals or make the cheese is relatively high, it is vital that each individual is fully occupied but not overworked. Failure can result from overwork as well as from inefficient work.

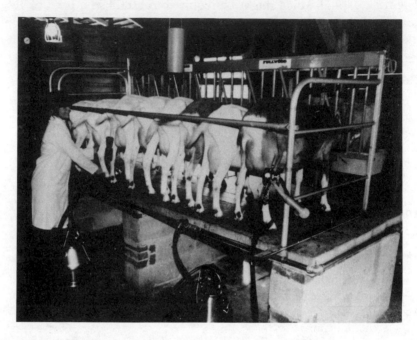

Plate 14 A simple milking parlour for smaller commercial dairy herds. (*Courtesy of Fullwood and Bland Ltd.*)

It is worth taking advice from the manufacturers of dairy equipment on how many goats can be milked per person per day through the various types of equipment. In this way it should be possible to match the capacity of the equipment to the herd so as to achieve maximum efficiency of work by the milking staff.

It is crucial to ensure that relief milkers are available to support the regular staff. In addition, time must be set aside, or bought, for administrative work and for delivery of milk or milk products to the customer.

Technical support is also essential. An independent expert can bring new knowledge and new approaches to the business, as well as providing technical and financial assessments of current activities.

Other fixed costs

Other fixed costs include charges for electricity, water, insurance, repairs to machinery, buildings and equipment, and rent or mortgage payments.

Table 9.6   Projected gross margins (£) for milk or cheese production

		Liquid milk	Cheese
Sales per head			
milk, 940 l[1]	@ £0.35/l	329	
cheese[2]: hard, 49 kg	@ 6.50/kg[3]	–	319
soft, 61 kg	@ £5.00/kg	–	305
culled females, 0.20	@ £25.00/head	5	5
kids, 10 days old, 1.75	@ £4.00/head	7	7
	Gross returns	341	636
Less herd replacements and mortality		19	19
	Output	322	617
Variable costs per head			
silage (low quality, from Table 9.4)		52	52
concentrates[4]		63	63
breeding males[5]		2	2
bedding		1	1
herd health scheme		15	15
disinfectant		4	4
cheesemaking consumables		–	30
Total variable costs		137	167
Gross margin per head		**185**	**450**

(1) 80% adults @ 1000 litres/head/year
    20% goatlings @ 700 litres/head/year
(2) Yields: hard, 10.5 kg/100 kg milk
            soft, 13 kg/100 kg milk
(3) From Table 9.1
(4) From Table 9.4 plus milk substitute
(5) Three males per 100 females @ £70/head

Margins

Gross margins

Projected gross margins for liquid milk or cheese, sold wholesale, are given in Table 9.6. It is assumed in the case of cheese production that half the milk output is processed into hard cheese and half into soft cheese. Mortality is assumed to be 5 per cent.

If all the milk output is processed into cheese, output is increased by 92 per cent and gross margin per head is increased by 143 per cent. But the increased gross margin does not take into account the extra labour involved in cheesemaking.

Net income

The gross margin method of economic analysis is most useful when comparing different enterprises on the same farm, where fixed costs are shared and difficult to apportion between individual enterprises. By contrast, a common situation in goat production is where the herd is the sole enterprise on a small farm. In some cases the holding has very little land but adequate buildings. In this case gross margin analysis gives little indication of net income since it does not account for fixed costs.

In Table 9.7 projected net incomes for liquid milk or cheese production are illustrated for a 100-head herd.

Since feed is costed at purchased price, the charge for rent or mortgage payments is for buildings alone. The alternative approach would be to enter home-produced feed in the gross margin analysis at cost of production and to include rent for the land used to produce the feed in the fixed costs.

The charge for labour includes casual labour for relief milking and administrative support including accountant's fees. It is assumed that all capital items are purchased with borrowed money, and that the loan is repaid over ten years at 16 per cent interest.

Whilst gross margins look reasonably attractive, only cheese production leaves a net income to prove a return on the capital invested in the business, to finance further investment to expand the business or to develop other enterprises.

Another way of viewing the relative net incomes in Table 9.7 is that a 100-head herd is the minimum size necessary to support one full-time

Table 9.7 Projected net income (£) for milk or cheese production, 100-head herd

	Liquid milk	Cheese
Gross margin (from Table 9.6)	18 500	45 000
less fixed costs		
rent or mortgage payments[1]	1 000	1 000
electricity and water	1 200	1 500
insurance	150	150
transport (delivery)[2]	1 500	2 000
labour	10 000	20 000
repairs and maintenance	700	700
capital items: annual charge[3]	3 700	4 500
Total fixed costs	18 250	29 850
Net income	**250**	**15 150**

(1) Buildings only
(2) Weekly delivery of frozen milk; less frequent deliveries of cheese over longer distances
(3) Annual charge: repayment of capital and interest at 16% on a 10-year loan of £18 000 for liquid milk or £22 000 for cheese production

time person, if liquid milk is the sole product. Thus either a larger herd is necessary to generate a reasonable return on capital invested, or a proportion of the output should be processed into cheese to add value and generate increased margins.

Further reading

Devendra, C. (1975) 'Biological efficiency of milk production in dairy goats.' *World Review of Animal Production*, **11**: 46–53.

Hunter, A.C., and Cruickshank, E.G. (1984) 'Hygienic aspects of goat milk production in Scotland.' *Dairy and Food Sanitation*, **4**: 212–215.

Jenness, R. (1980) 'Composition and characteristics of goats milk: Review 1968–1979.' *Journal of Dairy Science*, **63**: 1605–1630.

MAFF/ADAS (1982–86) Various leaflets on milk production and milk products.

The Milk and Dairies (General) Regulations. *Statutory Instruments*, No. 277, 1959 London: HMSO.

The Milk (Special Designation) Regulations. *Statutory Instruments*, No. 1542, 1960 London: HMSO.

Mowlem, A., and McKinnon, C.H. (1983) 'The production of high quality goat milk for retail sales.' in: *The Proceedings of the 3rd International Symposium on Machine Milking Small Ruminants*, Valladolid, Spain.

Sands, M., and McDowell, R.E. (1978) *The Potential of the Goat for Milk Production in the Tropics*. Cornell International Agriculture Mimeo Department of Animal Science, Cornell University, Ithaca, New York.

10 Profitable Meat Production

Goat meat, a by-product of the dairy industry, principally comprises the production from male kids and from female kids which are not required as herd replacements. Many dairy units rear their own kids for meat, but some sell their kids, either at a few days of age or after weaning, for rearing by specialist meat producers.

Markets

The principal markets for goat meat are in those countries where the goat population is relatively large and where goat meat is a common ingredient of the diet. In some European countries there are also seasonal markets for young milk-fed kids – for example Capretto in Italy and Chevrette in France. In other countries such as the UK there are opportunities for marketing locally to ethnic groups which are not only familiar with goat meat, but wish to eat it in preference to other meat.

Prices for goat meat are very variable and, as with goat milk, are subject to seasonal variations. Milk-fed goats command a price premium, especially at the time of religious festivals such as Easter or the end of Ramadan.

Goat meat is a low-fat meat. There are clearly opportunities to exploit this feature when marketing goat meat in areas where the population is aware of the need to reduce the consumption of fat in the diet.

Systems of production

A common system of goat meat production in many countries of Europe, Africa and Asia is to allow the kids to suckle their mothers until they are slaughtered at between 6 and 12 weeks of age. In some areas kids are given milk replacer *ad libitum* instead of whole milk. The meat from these animals is very lean and tender, and has a pale colour similar to that of young milk-fed lamb.

Other systems for rearing kids for meat involve weaning at about 8 weeks of age and thereafter offering concentrates *ad libitum* or concentrates with limited forage. Where concentrates are available *ad libitum* it is good practice to ensure that a small amount of hay, about 10 to 15 per cent of the ration, is also on offer to maintain rumen function. Water should be available at all times and both feed and water should be protected from contamination. A mineral supplement should be provided, preferably in the feed, but high intakes of magnesium or phosphorus should be avoided to reduce the risk of urinary calculi. The protein requirements of kids reared for meat have not been well-defined; suggested requirements are in Chapter 8.

Target rates of growth and target slaughter weights for the different systems are given in Fig. 10.1. For the all-milk system the animals should reach a slaughter weight of 15 kg at 7 to 8 weeks of age. The target slaughter weight of 30 kg should be reached at 18 to 19 weeks of age on the all-concentrate system and at 23 to 24 weeks of age on the semi-intensive forage/concentrate system.

The key to success is to maintain the stock in a healthy condition, and to have fresh feed available at all times. It is good practice to weigh the animals to check that the target growth rates are being achieved.

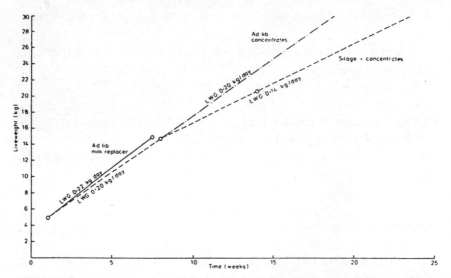

Fig. 10.1 Target growth rates of kids reared on milk replacer, concentrates or silage and concentrates.

Output

The income from goat meat production is determined by the weight of the carcass and its saleable meat yield. If the meat is sold retail, then the relative yield of different joints and the proportion of fat in the carcass may also influence the total return to the producer.

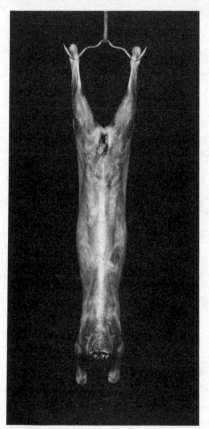 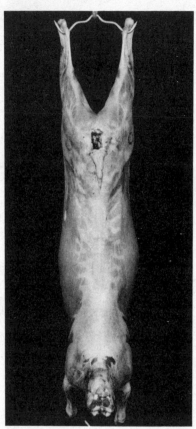

Plates 15 and 16 Castrated male Saanen kids slaughtered to produce carcasses weighing 9.6kg (Plate 15) and 24kg (Plate 16). Note the lack of subcutaneous fat, even in the heavier carcass. (*Courtesy of Food Research Institute, Bristol.*)

As the animal increases in liveweight there is a corresponding increase in carcass weight. Carcass yield is normally between 45 and 52 per cent of liveweight (Fig. 10.2). The lower values for killing-out percentage are more likely to occur in older animals given diets relatively high in forage. On the other hand, killing-out percentage commonly exceeds 50 per cent and may be as high as 60 per cent in milk-fed kids.

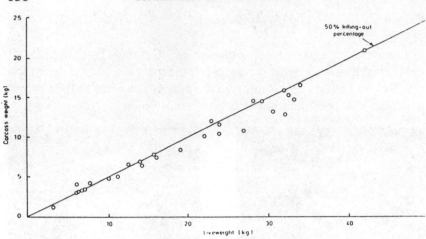

Fig. 10.2 Relationship between carcass weight and liveweight of kids reared for meat on all concentrate or concentrate plus silage systems.

A unique feature of the goat is the low proportion of subcutaneous fat in the carcass. Thus saleable meat yield is almost 100 per cent of carcass weight in goats slaughtered at 30 kg liveweight or less. Even in heavier animals, total fat in the carcass is relatively low, especially in entire males (Table 10.1).

Recent studies of the distribution of joints in female and castrated male kids slaughtered at 25 kg average liveweight show that the shoulder is the heaviest joint. Together with the leg the two joints

Table 10.1   Carcass composition of entire and castrated male goats

	Liveweight at slaughter (kg)					
	8.1		24.5		36.5	
	Entire	Castrated	Entire	Castrated	Entire	Castrated
Carcass weight (kg)	5.2	5.2	11.7	12.1	19.1	19.1
Carcass composition (%)						
fat	7.8	9.2	9.9	12.9	12.2	17.6
lean	55.4	56.1	58.3	58.6	61.2	54.2
bone	22.1	20.6	19.2	17.7	16.2	14.8

Source: Mtenga, L.A. (1979) *Meat production from Saanen Goats: Growth and Development* PhD Thesis, University of Reading

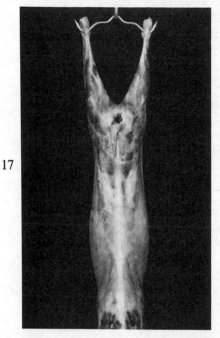

17

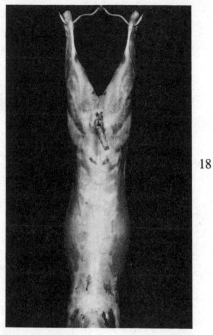

18

Plates 17 and 18 Saanen (17) and Angora x Saanen (18) castrated males slaughtered at 14kg carcass weight. The Angora x Saanen has a conformation which is more like that of lamb. (*Courtesy of Food Research Institute, Bristol.*)

account for 70 per cent of carcass weight (Table 10.2). The individual joints account for similar proportions of carcass weight over a wide range of liveweight at slaughter (14 to 46 kg).

Table 10.2 Distribution of joints in the goat carcass[1]

Joint	Percentage of carcass weight
Shoulder	39
Leg	32
Breast	12
Loin	10
Neck	5
Flank	2

(1) 12.4 kg average carcass weight

Source Butler-Hogg, B.W. and Mowlem, A. (1985) 'Carcase quality in Britain Saanen goats' *Animal Production* **40**: 572

The lean meat (muscle) content of the carcass averages about 68 per cent, though this value may vary from 64 per cent to over 80 per cent, depending on breed of animal and weight at slaughter.

In addition to the carcass, the skin constitutes an additional item of output. Some producers are able to secure significant revenues for their skins where outlets exist for leather or rug production.

Variable costs

The main variable cost in meat production is feed. Example feed budgets for the systems described earlier are given in Table 10.3.

Table 10.3 Feed budgets (per kid from birth to slaughter) for different systems of goat meat production

	System					
	Milk-fed[1]		High concentrates[2]		Silage/ concentrates[3]	
Targets						
slaughter weight (kg)	15		30		30	
weeks to slaughter	7.5		19		24	
Feed	(kg)	(£)[4]	(kg)[5]	(£)[4]	(kg)[5]	(£)[4]
milk replacer	14	11.2	11	8.8	11	8.8
concentrate (16% protein; 12.5 MJ ME/kg DM)	–	–	90	13.5	50	7.5
hay	–	–	12	0.6	–	–
silage (25% DM; 10.3 MJ ME/kg DM; 90 g DCP/kg DM)	–	–	–	–	300	5.4
total feed costs		11.2		22.9		21.7

(1) Week 1: colostrum and transfer to milk replacer. Weeks 2 to 8 (slaughter): milk replacer *ad lib*, average consumption 290 g milk powder/day, average liveweight gain 220 g/day

(2) Week 1: as (1). Weeks 2 to 8: restricted milk replacer and weaning completed by end of week 8; concentrates offered from 2 weeks. Weeks 9 to 19: concentrates *ad lib*, with small amount of hay; average liveweight gain 200 g/day pre- and post-weaning

(3) Diets: Week 1 and weeks 2 to 8: as (2). Weeks 9 to 24: restricted concentrates plus silage *ad lib*; average liveweight gain 200 g/day pre-weaning, 140 g/day post-weaning

(4) Feed costs per tonne: milk replacer £800, concentrates £150, hay £50, silage £18

(5) Wastage: concentrates 5%, silage 25%

Despite the relatively high cost of milk replacer, total feed costs are lowest for the all-milk system and highest for the all-concentrate system.

Other variable costs include those for bedding, transport, veterinary treatments, slaughter and butchery.

In total, the cost of rearing kids to slaughter should not exceed £30 per head.

Margins

Projected gross margins for the three contrasting systems of meat production are shown in Table 10.4. It has been assumed that the meat enterprise is operated as an additional enterprise on a dairy unit, using home-bred kids, and that the meat is sold retail.

Table 10.4   Projected gross margins for goat meat production

	System		
	Milk	Concentrate	Silage/ concentrate
Output per head (£)			
8 kg carcass @ £4/kg	32.0	–	–
14.5 kg carcass @ £2.75/kg	–	39.9	39.9
Variable costs per head (£)			
feed (Table 10.3)	11.2	22.9	21.7
bedding	0.1	0.3	0.3
veterinary	0.4	0.5	0.5
transport and miscellaneous	1.0	1.0	1.0
slaughter and butchery	5.5	5.5	5.5
Total	18.2	30.2	29.0
Gross margin per head	**13.6**	**9.7**	**10.9**

If weaned kids are purchased for the concentrate or silage/ concentrate system, feed costs will be lower (by £9 per head) but the animals are likely to cost between £10 and £15 per head, depending on their age and weight.

Margins may be increased by adding value to the meat – either through novel methods of cutting and presenting the meat to the consumer, or through processing the meat into sausages, pâté and other meat products.

Further reading

Devendra, C., and Owen, J.E. (1983) 'Quantitative and qualitative aspects of meat production from goats.' *World Animal Review*, **47**: 19–29.

McDowell, R.E., and Bove, L. (1977) *The Goat as a Producer of Meat.* Cornell International Agriculture Mimeo No. 56 Department of Animal Science, Cornell University, Ithaca, New York.

Owen, E., and Mtenga, L.A. (1980) 'Effect of weight, castration and diet on growth performance and carcass composition of British Saanen goats.' *Animal Production*, **30**: 479.

Owen, E., and De Paiva, P. (1983) 'Artificial rearing of goat kids; effect of age at weaning and milk substitute restriction on performance to slaughter weight.' *Animal Production*, **30**: 480.

Stark, B.A., and Wilkinson, J.M. (1985) 'The growth of goats given contrasting supplements to hay.' *Animal Production*, **40**: 572.

11 Profitable fibre production

The proportion of the world goat population which produces mohair and cashmere is relatively small, but the value of the products is high and demand for both products is strong as consumers increasingly seek natural textiles of high quality.

Most of the mohair produced in the world is from Angora goats kept in the more arid regions of South Africa, Turkey and the USA, whilst most of the world's cashmere production originates in China.

Markets

A high proportion of the world production is imported into the UK for spinning into textiles (Table 11.1). This represents an extremely lucrative potential market for producers in the UK and elsewhere in Europe, where production of both fibres is very small indeed.

The major processors of cashmere, Dawson International Ltd, in Scotland, are keen to develop the production of cashmere in the UK, Australia and New Zealand. It is possible that similar opportunities exist for mohair production. In both cases it is likely that production will be organised through groups of producers who together can provide the textile manufacturers with relatively large quantities of fibre at each delivery.

Table 11.1 Importations of mohair and cashmere into the UK, 1985

	Quantity imported (tonnes)	Value (£m)
Mohair	7356	46.1
Cashmere	2188	55.9

Source: HM Customs and Excise Statistical Office, personal communication 1986

Skins may also be cured for sale as rugs. This product is more suited to dairy breeds which do not produce mohair or cashmere. Most of the goatskin rugs at present originate in Asia.

An alternative market for skins from young animals is for processing into high quality kid leather. Some tanneries are interested in obtaining bulk supplies of kid skins, but the market for this product is not well established in the UK.

Prices

Typical whosesale prices for Angora kids, mohair, cashmere and goatskin rugs in the UK are given in Table 11.2. At the time of writing (1986) there is a very strong demand for female Angora and Angora × Saanen kids, reflecting a high level of interest amongst farmers wishing to develop fibre production from goats.

Table 11.2 Typical UK prices for female Angora kids, mohair, cashmere and goatskin rugs, 1986

	Typical price
Angora kids:	
purebred females, minimum 6 months old (£)	4500
C grade[1], minimum 6 months old (£)	900
Angora × Saanen kids: female, minimum 6 months old (£)	300
Mohair: wholesale (£/kg)	8 to 10
Cashmere: wholesale, top quality (£/kg)	50 to 60
Goatskin rugs: Chinese (£)	25 to 30

(1) ¾ purebred Angora

Systems of production

Traditionally both Angora and cashmere goats are grazed extensively on a low-input–low-output system of management. Consequently, reproductive performance is relatively poor and yields of fibre are variable. Cashmere is often a secondary product from goats kept principally for meat or milk, whereas mohair is usually a primary product with meat as a by-product. In some areas, such as South Africa and the USA, Angora goats are kept under lowland conditions where pasture yields are higher and herd output is greater.

Recently, interest has developed in keeping Angora and Angora × Saanen goats in temperate areas with intensive pasture management and with housing during periods of low grass production or inclement weather. The objective is to develop a fibre enterprise as an alternative livestock enterprise.

Output

The main characteristics of economic importance in fibre production are the weight of clean fibre, fibre fineness, staple length, type of lock (mohair), and proportion of kemp (mohair). In addition, an acceptable level of reproductive performance, a low mortality rate, and low coat loss due to stress, all influence the output of the herd.

Fibre growth and yield

Skin follicles develop during pre-natal growth of the foetus. Until the 12th week of pregnancy they are primary (hair) follicles; thereafter secondary (wool) follicles appear. At birth the ratio of secondary to primary follicles (the S:P ratio) is about 7:1 and the ratio of secondary fibre to primary fibre about 1.5:1. Post-natal development consists of new secondary follicles so that the S:P ratio in Angora goats increases to 8.6:1 at three months of age, and to 9:1 at four months of age.

Cashmere goats have a double coat with a coarse outer coat of primary fibres, known as kemps, and a thin, fine undercoat of secondary wool fibres. In Angoras with kemp the proportion of kemp decreases with age from about 44 per cent at birth to 7 per cent at three months of age. Thus most adult Angoras produce a single coat of long, coarse but non-hairy fibres.

Mohair consists of the protein keratin, but it has a characteristic lustre which most wool does not possess, and a higher tensile strength.

Of the factors affecting the quantity of fleece produced per year, age, season and nutrition are the most important. Mohair production increases from birth and peaks at three to four years of age. Data from recorded herds in Texas, USA, show slower fleece growth in winter

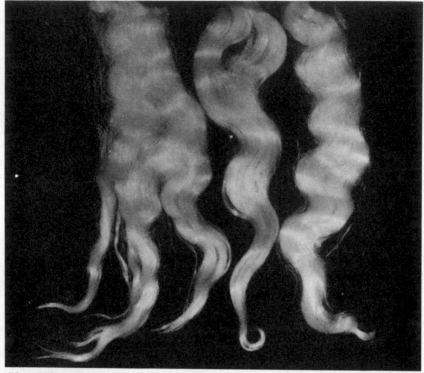

Plate 19 Samples of mohair from Angora goats. From left to right: adult; young goat; kid mohair. (*Courtesy of International Mohair Association.*)

Plate 20 An adult feral male used for the production of cashmere fibre. (*Courtesy of Hill Farming Research Organisation.*)

than in summer so that staple length is about 2 cm shorter in animals shorn in spring than in those shorn in the autumn (Fig. 11.1). However, seasonal differences in weight of fleece appear to be less in animals of more than three years of age than in younger animals.

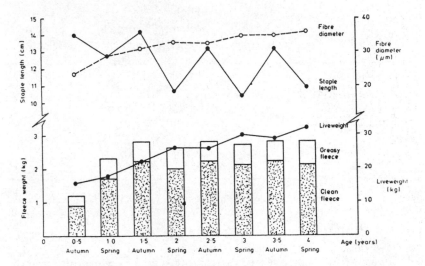

Fig. 11.1 Liveweight, fleece weight, staple length and fibre diameter of Angora goats in Texas shorn in autumn and spring. Source: Shelton, M. (1981) Fiber production *in Goat Production* (Ed. C. Gall) London: Academic Press.

Yield of mohair and cashmere is generally 75–80 per cent and 60 per cent, respectively, of greasy fleece weight. The lower yield for cashmere reflects the need to separate cashmere fibres from hair from primary follicles.

Nutrition can exert a marked influence on rate of fleece growth. Mohair production in the Angora goat, like wool production in the merino, is particularly sensitive to protein supply. A study in South Africa (Table 11.3) demonstrated that the level of energy and protein in the diet of the dam influenced both the birth weight of the kid and its weight at five months of age. But the yield of fleece from the kid was reduced by only 6 per cent on the low energy diets as compared to the high energy diets, whilst fleece yield was reduced by 29 per cent on the low protein diet as compared to the high protein diets.

Targets for yield of mohair and cashmere are given in Table 11.4. They relate to adult animals. In the case of mohair it is assumed that the goats are sheared twice per year.

Table 11.3 Effect of undernutrition of the dam on growth and fibre production of Angora kids

	Diet of dam			
	High protein, high energy	High protein, low energy	Low protein, high energy	Low protein, low energy
Birth weight (kg)	2.9	2.4	2.4	2.2
Weight at 5 months of age (kg)	20.9	18.7	17.7	14.9
Weight of greasy fleece at 5 months of age (kg)	1.23	1.13	0.87	0.84

Source: Van der Westhuysen, J.M., Wentzel, D. and Grobler, M.C. (1981) *Angora Goats and Mohair in South Africa* S. African Mohair Grower's Association, Port Elizabeth, South Africa, p. 98

Mohair grows at an average rate of 2.0 to 2.5 cm per month and the length of the fleece is primarily a reflection of the inter-shearing period. The price for mohair fleece is influenced very little by length of fibre, provided it is longer than 7.5 cm. Fleeces shorter than 7.5 cm are likely to realise about 80 per cent of the price of longer fleeces. Thus the interval between shearings should be at least four to five months, and there is little justification for shearing more than twice a year.

Cashmere fibres are traditionally combed out of the fleece as they are shed in the spring, but in some areas where yields are higher they are harvested by shearing. After harvest the product is sorted into grades based on fineness, colour and proportion of down.

Table 11.4 Target production of mohair and cashmere

	Mohair	Cashmere
Greasy fleece (kg/year)	6.0 (2 shearings)	0.5 (combing)
Clean fibre (kg/year)	4.8	0.4 (dehaired)

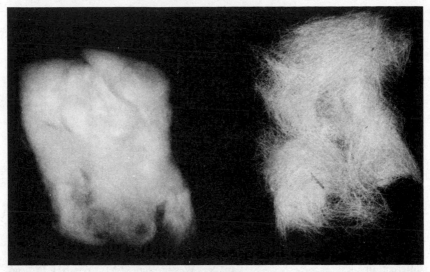

Plate 21 Samples of cashmère down (left) and hair (right). The down is the valuable undercoat which is used for the manufacture of luxury textile products. (*Courtesy of Dawson International Ltd.*)

Fibre quality

Fibre diameter is the most important attribute of fibre quality in both mohair and cashmere and has a major influence on the price realised for the products.

Fineness of mohair fibre decreases with age, averaging less than 25 μm (microns) at the first shearing to over 35 μm at four years of age (Fig. 11.1). Kid fibre, although of the highest quality, only represents a small proportion (about 15 per cent) of the total weight of fleece from the herd.

Adult Angoras commonly produce mohair of between 36 and 46 μm diameter, and the price for fibre of this quality averaged less than 60 per cent of that for the finest (kid) quality fibre over a 10-year period in South Africa (Fig. 11.2). Mohair of high quality is soft but firm to the touch, has a bright lustre and forms ringlets of wavy, solid-twisted staples of uniform fibre length. Any foreign matter, kemp or coloured fibres will decrease the value of the product, as they must be removed during processing. Urine, some soils and vegetable matter can stain the fibre permanently and it is often necessary to remove stained areas.

Cashmere is considerably finer and shorter than mohair. The superior (knitwear) grades are less than 15.5 μm in diameter. Weaving

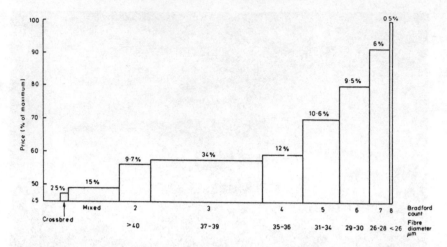

Fig. 11.2 Price and percentages of the South African mohair clip over 10 seasons in different fineness classes. Source: Van der Westhuysen, J.M., Wenizel D. and Grobler, M.C. (1981) *Angora Goats and Mohair in South Africa* Port Elizabeth, South Africa: South African Mohair Growers Association.

grades average 18 μm in diameter. The length of cashmere is usually less than 6 cm. White is preferred to coloured fibre, but natural colours are used in the production of some cashmere garments.

Kids

Poor reproductive efficiency is a major problem with the Angora. Apart from limiting total herd output, a low rate of reproduction reduces the selection pressure in the herd and slows down the rate of genetic gain. Also, surplus female kids are slaughtered for meat production and thus the annual output of meat is influenced by the total number of kids born per year.

Fibre production tends to take precedence over growth, reproduction and milk output in Angoras. Small females have lower ovulation rates, and lower conception rates than larger, heavier animals. Stress, caused by low temperatures, inadequate nutrition or transportation, can also result in abortions. The effect of liveweight at breeding on kidding percentage is shown in Table 11.5.

Abortions tend to be noticed when they occur in mid-pregnancy, but there is evidence from studies in sheep that inadequate nutrition at breeding time can be reflected in an increase in early embryo loss. In

Table 11.5 Effect of liveweight of Angora female goats on kidding percentage

Liveweight at breeding (kg)	Kidding percentage (kids per 100 does to the buck)
<23	50
23 to 27	74
28 to 31	68
32 to 36	93
37 to 40	105
41 to 45	167

Source: Shelton, M. (1981) 'Fibre production' in: Gall, C. (Ed) *Goat Production*, London: Academic Press, pp 379–409

addition, females should be gaining in weight at breeding time to ensure a high ovulation rate.

Loss of kids can occur shortly after birth if kidding occurs under extensive range conditions. Common causes of loss are cold stress, starvation and predation. An additional problem is poor mothering, when the mother abandons her offspring. It is essential that, if at all possible, females are confined at kidding, either indoors or in small paddocks. Adequate nutrition and as little stress as possible can result in relatively low post-natal mortality.

There is some evidence that poor reproductive performance, especially abortion, is linked to the level of mohair production relative to body size. Thus selection for increased fibre output should also take account of the need to maintain reproductive efficiency. In practice this means that selection should be for increased liveweight.

Further reading

International Mohair Association. *Information sheets*.

Ryder, M.L. (1986) 'Progress with British cashmere production.' *Wool Record*, April 1986 p52.

Ryder, M.L. (1986) 'High quality textile fibres from goats.' *Span*, **29**: 29–31.

Ryder, M.L. (1986) 'The goat.' *Biologist*, **33**: 131–139.

Van Der Westhuysen, J.M., Wentzel, D., and Grobler, M.C. (1981) *Angora Goats and Mohair in South Africa*. Port Elizabeth, South Africa: South African Mohair Growers Association.

12   A Triple-Purpose Goat?

The theme of this book has been that goats can be farmed commercially and profitably like other ruminants, provided that the scale of operation is adequate and that technical efficiency is of a sufficient standard to ensure that the stock are healthy and productive.

Specialisation

It follows that specialisation is necessary so that appropriate attention can be given to marketing, product quality, output and costs of production. Thus the goat industry has developed to the point where some larger dairy goat farmers sell liquid milk to specialist processors, who sell through distributors to supermarket and health food chains. This pattern is the same as that for cow milk. Other producers employ staff to process and package their milk and to add shelf life, value and originality to the product.

With the establishment of stabilised and organised markets for milk, confidence is growing amongst dairy goat farmers. But similar markets are not yet well established for goat meat, though the emergence of a few large meat producers and meat wholesalers may well help to generate a similar degree of confidence.

In the case of fibre production, the issue is one of supply. Angora and cashmere goats are at present relatively scarce in temperate areas of Europe, America and Oceania. Continued importations of animals and embryos into Europe will occur, and lead to the establishment of large-scale commercial producers who will form the core of the European goat fibre industry. Fortunately the market for fibre is already very well organised and well established. For example, Dawson International Ltd currently process about half the total world production of cashmere.

The goat is an old-established and genetically diverse species of domesticated animal. This is fortunate because it means that the genes for particular attributes such as fine fibre, high meat yield or high milk solids can be readily identified.

Conventional breeding has indicated that several useful genes are not correlated. Thus the scope for breeding for both high milk yield and good carcass conformation may be very low. Similarly the correlation between reproductive performance and fibre yield appears to be negative. It may also prove to be futile to cross Angoras with Saanens in an attempt to produce animals with both high milk production and high fibre yields.

Diversification

It appears that specialisation will continue in commercial goat production – at least in the short term. But some trends in diversification can already be identified, which may develop into large-scale activities.

Some dairy goat farmers, seeing the business opportunity offered by mohair production, have started to cross pure Angora males with their Saanen-type females. Repeated back-crossing to the Angora gives a progressive increase in fibre yield and in fibre quality, so that four or five generations later the animals are fully 'graded-up' to the Angora type. This diversification has already resulted in a useful supplementary income, in addition to that from milk, from the sale of crossbred females at inflated prices (see Table 11.2, page 144). Fortunately the male kids have a superior carcass conformation for meat production to that of the pure Saanen.

By contrast, the realisation that feral goats can yield cashmere and also produce meat has prompted the formation, in June 1986, of the Scottish Cashmere Producers' Association, which has announced plans to establish herds on hill farms where sheep and beef cattle have traditionally been grazed. This may mean that beef and sheep farmers are encouraged to diversify into goats. Apart from the benefit of the additional income from cashmere, the grazing habits of the goat mean that they can improve hill pastures for grazing sheep and cattle (see Chapter 8).

A triple-purpose goat?

The advent of gene transfer in animals has profound consequences for commercial goat production. It is already technically possible to introduce foreign genes into animals and to achieve their expression – if necessary in specific organs of the body. It now remains for the genes for attributes of economic value to be fully identified. They may then be transferred to the appropriate breeds of goat to change their performance characteristics. In other words, it should soon be possible to engineer a truly triple-purpose goat – one with a high milk yield, a high yield of mohair or cashmere, *and* a high meat yield. This achievement would completely change the strategy of commercial goat production. Commercial units would be developed to exploit the uniquely high value of the output from the animal, and the people who produced the new 'breed' would make a very large amount of money.

In the meantime, the wisest approach is to concentrate on improving the efficiency of producing one main product. If individual farm circumstances permit diversification, so much the better.

Either way, the future looks good for commercial goat production.

Further Reading

General

Baker, F.H. (Ed) (1983) *Sheep and Goat Handbook* Volume 3 International Stockman's School Handbooks, West View Press, USA.

British Goat Society *Yearbooks* and *monthly journals*.

Dairy Goats, Milk and Milk Products Seminar, Papers from a Seminar, 4 November 1982 Cheshire College of Agriculture, Reaseheath, Cheshire.

Devendra, C. and Burns, M. (1982) *Goat Production in the Tropics* Farnham Royal: Commonwealth Agriculture Bureaux.

Gall, C. (Ed) (1981) *Goat Production* London: Academic Press.

Goat Producers Association 1984, 1985 *Developments in Goat Production* (Eds J.M Wilkinson and A. Mowlem).

Goat Veterinary Society *Quarterly Jounals*.

Haenlein, G.R.W. (1981) 'Dairy goat industry of the United States' *Journal of Dairy Science* **64**, 1288–1304.

Hetherington, L. (1977) *All About Goats* Ipswich: Farming Press Ltd.

Islay and Jura Goat Society (1985) *Goat Husbandry Survey Report 1985* ICSS, Highfield, High Street, Bowmore, Isle of Islay, Argyll.

La Chèvre *Monthly journal* ITOVIC, Paris.

International Goat and Sheep Research *Journals* Scottsdale, Arizona; Dairy Goat Journal Publishing Company.

McFarland, A.G. (1982) *Goat Farming* MAFF/ADAS.

MacKenzie, D. (1980) *Goat Husbandry* (4th edition revised and edited by J. Laing) London: Faber and Faber.

Mottram, T. (1985) *The Viability of Commercial Goat Farming in the United Kingdom* Middle Tremollett, Coads Green, Launceston, Cornwall.

National Institute for Research in Dairying *Annual reports* 1980–1985.

Proceedings of the International Symposium on Nutrition and Systems of Goat Feeding Tours, France (1981) Paris: ITOVIC.

Proceedings of the Third International Conference on Goat Production and Disease Tucson, Arizona (1982) Scottsdale, Arizona: Dairy Goat Publishing Co.

Quittet, E. (Ed) (1975) *La Chèvre* Paris: La Maison Rustique.

Roberts, D. (1985) 'Microbiological aspects of goats milk: Public health laboratory service survey' *Journal of Hygiene Cambridge* **94**: 31–44.

Sands, M. and McDowell, R.E. (1978) *The Potential of the Goat for Milk Production in the Tropics* New York: Department of Animal Science, Cornell University.

Scott, S.A. (1984) *Commercial Goat Dairying* MAFF/ADAS.

Scott, S.A. (1985) *Report on Commercial Goat Farming in France* MAFF/ADAS.

Smallholder and Goatkeeper *Monthly magazine.*

Thornhill, S. (1984) *Commercial Dairy Goat Farming in the UK* Thesis, Department of Agriculture, University of Reading.

Vincent, R.J. (1983) *Commercial Goat Farming* 'Report on a visit to Australia, New Zealand and USA in June to August 1982 to study the opportunities for UK Dairy Goat Farming' Olney, Bucks: Nuffield Farming Scholarships Trust.

Wilkinson, J.M. and Stark, B.A. (1982) *Goat Production: Some Possibilities for Goat Production in the UK* MAFF.

Index